10

FOR T ...NG DESIGN AND TECHNOLOGY

David Spendlove

continuum

Continuum International Publishing Group

The Tower Building	80 Maiden Lane
11 York Road	Suite 704
London, SE1 7NX	New York, NY 10038

www.continuumbooks.com

First published 2008
Reprinted 2009, 2010, 2011

British Library Cataloguing-in-Publication Data
A catalogue record for this book is available from the British
Library.

ISBN: 9780826499752 (paperback)

Library of Congress Cataloging-in-Publication Data
Spendlove, David.
    100 ideas for teaching design and technology / David Spendlove.
        p. cm.
    Includes bibliographical references.
    ISBN-13: 978-0-8264-9975-2 (pbk.)
    ISBN-10: 0-8264-9975-9 (pbk.)
    1. Design, Industrial—Study and teaching (Elementary). 2.
Design, Industrial—Studay and teaching (Secondary). 3.
Technology—Study and teaching (Elementary). 4. Technology—
Study and teaching (Secondary) I. Title. II. Title: Hundred ideas
for teaching design and technology.

    TS171.4.S66 2008
    372.35'8—dc22

Designed and typeset by Ben Cracknell Studios
www.benstudios.co.uk

Printed in India by Replika Press Pvt. Ltd.

100 IDEAS
FOR TEACHING DESIGN
AND TECHNOLOGY

CONTINUUM ONE HUNDREDS SERIES

CONTENTS

SECTION 2 **Designing**

SECTION 3 **Using technology**

SECTION 7 Assessment

SECTION 8 The wider classroom

INTRODUCTION

This book is very much about 'how' rather than 'what' to teach in design and technology (D&T); it draws upon best practice in teaching and locates this within a D&T context. In doing so it provides a mixture of ideas, some of which can be implemented straight away, while others provide opportunities for reflection. The key focus remains upon well-conceived and researched methods of improving learning through good teaching.

The need for such a focus is critical as much of the emphasis in the evolution of D&T as a subject has been on 'what' to teach, with a significant focus on resources and new technologies. Although resource development in the subject is important, by far the biggest and most important resource in any classroom is the teacher, and it is essential that you have both a clear philosophy about what you want to teach and a coherent repertoire of abilities to enable you to teach effectively.

However, knowing 'what' to teach is also important and goes hand-in-hand with 'how' to teach. To reinforce this point there is an age-old story about a new teacher who tried to explain a new topic to a group of students, but he got the impression that they didn't understand. Slightly frustrated, he then taught the topic in another way, but again they didn't understand. The teacher didn't want to retreat, and, growing increasingly frustrated, he tried delivering the topic a third way. Then he realized that actually he didn't understand it himself! The happy ending of this story is that the teacher, although focusing on the 'how', finally realized he didn't understand the 'what', and once he did, he was able to help the students to build their understanding. So although this book is about 'how' to teach, don't forget you have to know your 'what' as well!

The big picture

TEACHING AND LEARNING

Are you the kind of teacher that lights up children's faces when you walk into the classroom? Or are you the kind of teacher that, on leaving the classroom, children's faces light up!

An important point to make early on is that although this book is titled *100 Ideas for Teaching Design and Technology*, it is important to recognize that this also includes learning.

Why the distinction? There is an old Snoopy cartoon where Charlie Brown says that he has taught his dog to whistle. When asked to demonstrate this he replies that the dog hasn't learned yet. Although I often see a lot of teaching taking place, I don't always see the accompanying learning! I have seen some great, entertaining lessons where the children have been occupied and have found whatever they have been doing quite interesting, but when I ask them what they are learning they haven't got a clue.

In D&T one of the unique features is that children learn by doing, through taking action. However, if you just have the doing – without the learning – then you have merely a sweatshop mentality of 'making things'.

So, pupils might be making X, Y or Z, but what are they learning? Often this isn't sufficiently clear, and the pupils are expected to absorb the learning through a process of osmosis. If they do learn in this way, it can often not be what the teacher had intended.

Good learning has to be planned, constructed, signalled, scaffolded and signposted; without this we are merely occupying the middle ground between therapeutic basket-weaving and the exploitative child labour found in poorer countries.

Teaching and learning through doing and taking action offers a unique and powerful experience, but only if the ratios between the doing and the learning are appropriately balanced. Therefore next time you plan a learning activity (lesson), divide a page lengthways into two columns representing a third and two thirds of the width. The narrower column represents the teaching column and the wider column is the pupil learning column. Begin by completing, in detail, the learning column. Then afterwards complete the teaching column in less detail, asking yourself what is the best way to teach the learning that you have identified. You will be surprised at the difference this makes to your lessons!

The reasons why we teach D&T are not set in stone, and the subject remains something of a maverick activity as its strengths can also be its weakness; the breadth of activities within the subject mean that opinions about why we teach D&T are wide ranging.

When offering suggestions about why the subject is on the curriculum, we also have to be mindful of what parts of the subject are unique, otherwise how can its place on an overcrowded curriculum be justified?

Ultimately justification has to be personal, and my belief is that the subject offers children a unique form of empowerment and learning. In D&T children (should) learn how to take action through working creatively and collaboratively. They engage in processes that challenge them in high-level thinking and decision making, considering values and emotions through rich and stimulating contexts.

Through these processes and actions they become autonomous, discerning and informed by learning through doing: learning in a 'just in time' approach rather than a 'just in case' approach. The skills they learn are empowering, diverse and appropriate to their action-taking, but it is essential that all pupils are aware of the implications of any actions they take.

The rationale that you or I have, however, is insignificant unless it is shared – so in order to challenge any misconceptions, get pupils to write down and discuss why they should or shouldn't study D&T.

D&T is delivered through the context of the designed and manufactured world, and the decision-making and action-taking that learners engage with may result in products, systems or environmental outcomes. The production of such outcomes is merely the vehicle for engagement in the unique activities outlined so far. To pursue these product outcomes alone, however, will result in an impoverished experience, failing to engage with the unique contributions offered by the subject.

A key principle of the subject is that learners take ownership of their world and in doing so seek to improve the world they live in through enquiry and exploration. Such an aim is both powerful and complex, but facilitates the development of appropriate skills, attitudes, concepts and knowledge.

The benefits from such a rich process are to children culture, society and industry.

Often such strong rationales can get lost in the daily grind, therefore it is essential that the key priorities are built into each activity and these are used to test each activity that you have in your curriculum.

Write down your definition of, and key characteristics of, the subject, and then ask yourself if each activity lends itself to your description? If they don't – you know what to do!

SOMETHING TO TAKE HOME?

There is a great urban myth in D&T that children need to take home something that they have made in the subject. It is perpetuated from generation to generation, and has become a driver for the subject without any real questioning of why. Yet for some reason this doesn't happen in other subjects.

It is not a problem that children might, as a consequence of an activity, have an outcome from their endeavours; the problem arises when this shapes the lesson and becomes the *raison d'être*.

The production of various 'knick-knacks', 'nibbles' and 'apparel' represents, at times, a sweatshop mentality where the production of such unnecessary items and memorabilia ultimately drives the experience. The myth is so deeply ingrained in many schools that you can see the child been edged out of many activities by the teacher, and reduced to merely placing 'clip art' on a predetermined shape which will then be cut by a laser cutter while the child watches.

The truth is that during our lives we do not carry around a collection of goods that we have accumulated, showing each new friend 'what I made in D&T'. Any such products have a limited shelf life. It is the actions we take and the learning that accompanies this which shapes us and has the lasting effect.

The D&T teacher's challenge in the twenty-first century is a very different one, and is almost the antithesis of the 'take home' mentality of the past. Learners now need to be considering the environmental, emotional and sustainable impact of their decision making rather than the inappropriate production of unnecessary items.

A simple way forward is to discuss this at your next departmental meeting and consider how D&T should reconcile the tension of producing items within a sustainable and environmental agenda. It may be that you aim to reduce material consumption within your department (e.g. reduce consumption over the next three years by 25%) and you share this target with pupils so that they participate in achieving the target.

The great aim of education is not knowledge but action.

Herbert Spencer

D&T has had a chequered history. It grew out of a range of subject and initiatives and has often struggled to gain a real sense of identity; it remains the second youngest subject on the curriculum. To those who understand it, it offers so many opportunities but unfortunately at times it has taken significant effort to try and shake off some of the prejudice that has held back its development.

An important part of this is whether the subject is an academic or vocational subject and who should study it. As a gross generalization, the subject has often been taken by pupils who are not very good at the sciences, mathematics, English or the humanities. However, pupils who were good at art and design would often take the subject as at least they knew there were careers for them in design. In many ways, when D&T became a compulsory subject for all pupils, it hadn't developed sufficient expertise in dealing with the full ability range, and therefore struggled in some schools to convince parents of the value of the subject.

However D&T is unique, and offers all pupils who are capable in the sciences and humanities the opportunity to not only maintain their interest in these areas but also to very much develop their capabilities by drawing upon all that they know and applying it to real-life situations. D&T is therefore 'vocademic' in nature and requires the full ability range of learners to extend and apply their expertise from other activities, and as such it should be at the heart of the school and curriculum.

Activity: Consider what 'vocademic' means in your department by examining what opportunities exist for able pupils, through activities, careers advice and extra-curricular activities, to utilize both academic and vocational challenges and insights.

Einstein has been quoted as saying 'you can't play the violin if you have just been using a large hammer', and this is a useful metaphor for the use of 'starter' activities within D&T as part of Key Stage 3 strategy.

One of the dangers of 'starters' we have to avoid, however, is that all lessons don't become familiar and that the pupils don't go from one starter activity to another in different classrooms and think that it all seems rather similar – therefore it requires discernment from the teacher to ensure that starter activities are used in the correct way, at the right time, and are used in a proper and positive way.

A good planned starter will differentiate to meet the needs of pupils on different occasions. Not all lessons will need starters, and some lessons will need physical starters to wake pupils up, while some lessons will need visual starters to calm pupils down.

When thinking of starters we also need to consider different models of delivery as part of effective teaching. For instance, commonly poorly-conceived starters are those which reflect the notional three-part lesson. In reality D&T lessons are often multi-phased and trying to distil them into three parts may be damaging.

In a typical D&T lesson there is going to be a large amount of self-directed pupil activity following the initial starter. The starter therefore may be used as part of a behaviour management strategy in settling pupils down after break, waking them up on a warm day and getting them visually orientated towards a creative lesson.

Remember – it is really important that the starter activity, in most cases, should be contextualized and linked to the lesson that is about to take place, and that it orientates pupils towards the learning activity rather than being a stand alone activity.

Don't let the starter spoil the main course!

An alternative to the traditional starter activity is where the lesson has many different episodes, including regular plenaries, and each episode has a different starter activity each time it is re-orientated.

A further model is the use of a starter activity that may be presented at the start of the lesson but not answered until the end. This might be appropriate when dealing with a broad concept introduced at the beginning, for instance a big thinking question, which can continually be revisited throughout the lesson (for example, asking 'has anyone got the answer yet?'). Such a lesson could deal with a difficult concept which can be contextualized in the initial starter activity, setting the scene, and then revisited regularly, feeding back into the main body of the lesson.

WHY STARTERS? (2)

WHEN SHOULD I HAVE PLENARIES?

Plenaries should not be considered as the sugar coating of entertainment over the same old learning – they are integral to securing pupils learning.

One of the strongest features of the introduction of national strategies was that they unified language used in the classroom and extended teachers' vocabulary. However, in doing so they created misconceptions among teachers, most notably that every lesson consisted of three parts, beginning with a starter and ending with a plenary.

Although this is one way of viewing a lesson, the reality is that the average D&T lesson is much more complex than this and plenaries will take place not just once but throughout the lesson.

Therefore although a plenary may take place at the end of the lesson to monitor the extent to which the objectives of the lesson have been achieved, within a D&T lesson the monitoring and reorientation of pupils will take place regularly. In practical work this may take the form of adding layers of complexity through demonstrating different skills and processes, but prior to doing this a mini plenary would assess the extent to which pupils would be challenged by the next target. For example, in design this could be through monitoring design progress, followed by diversifying the stimulus while drawing attention to good practice.

A further misconception is that a plenary should be a fun game. Although it is desirable to have enjoyable activities, often the 'fun' element can override the learning and is part of the creeping 'edutainment' philosophy in education, where children are considered to learn best through 'fun', rather than gaining enjoyment from the process of learning.

Reminder: Each mini plenary provides a monitoring, assessment and reorientation opportunity, so to risk leaving this to the end of the lesson may be too late.

Glass ceiling or sticky floor?

How do you think about the National Curriculum in D&T? Is it a sticky-floor scenario, which just pulls you down and where the breadth of what needs to be covered prevents you from delivering effectively?

Or do you think of the National Curriculum as a glass ceiling – something that limits the overall scope of your pupils' development as you want to go beyond what is prescribed but feel it best to stick within the guidelines?

First, as in all situations, we have to look for the positives, and in many ways teachers often forget that, although not perfect, the National Curriculum has enshrined in law what D&T is, what it should look like and without it the momentum for the subject would be lost.

Second, there is something of a myth about how the National Curriculum should be delivered which has transcended the various formats of previous versions. So, early versions of the National Curriculum were a potential 'sticky floor' variety of over-prescription. However, the latest version is very much a loosely-controlling document which allows genuine freedom and encourages teacher autonomy.

One aspect that provides an incredibly rich orientation to D&T is the 'statement' which outlines the overall aims of the subject. This provides the very reason, the DNA, of why we have D&T – and it is where any curriculum developments should come from. If you have not read it recently, I would advise you to do so, and you should consider the extent to which you genuinely address all areas of the statement. Also, when considering the statement and the development of the curriculum, rather than asking 'to what extent do my activities fit aspects of the statement?', start from the other end and ask 'how best should each area of the statement be addressed?' You will be surprised at the difference!

IDEA

10

PRIORITIES (200 HOURS)

If you were told you had 24 hours to live, you wouldn't start to learn how to play the piano!

There is a fundamental flaw with any discussion about D&T, and that relates to the lack of priorities in the subject. For instance, there are those who still maintain that good old-fashioned skills are at the heart of the subject, while others may give priority to a creative approach, a highly technical approach or a combination of these – delivered through one or all of the different routes: design, food, textiles, graphics, engineering, materials, electronics, and so on.

Although these assertions are admirable, they are also often slightly naïve, as the reality is that developing real capability in any facet of D&T takes time and in an overcrowded and competitive curriculum this is something most of us don't have.

The bottom line is now that D&T is no longer compulsory at Key Stage 4, many children will only ever experience their minimum entitlement at Key Stages 1, 2 and 3. If they are fortunate enough to receive their entitlement then this amounts to little more than 200 hours. Within this time, the majority will have been delivered at Key Stages 1 and 2 (sometimes by non-specialists) while their Key Stage 3 experience may have been divided into as many as five different material areas. So their 200 hours looks even more diluted as an experience especially as it has taken place across nine years!

Ultimately any conception of D&T has to be prioritized in the context of the figures presented, bearing in mind it is considered that to be truly proficient at a subject you need around 10,000 hours study, which equates to 50 school lifetimes.

With this in mind a way of prioritizing an initial focus for teaching D&T is that a very small number of those we teach will go on to be designers, technologists, engineers, craftspeople, etc. But every one of them will be both a powerful consumer and a participatory citizen, where an insight into, and understanding of, issues of design, values, sustainability and taking action will be invaluable.

Activity: Calculate how many hours your pupils will experience in D&T and then make a list of priorities of how and what they should learn. Your task will then be to best plan how to deliver your priorities in the limited time that you have – not an easy task!

GLOBAL EDUCATION (1)

80% of the environmental impact of a product is determined at the design stage.

In his book *In the Bubble: Designing in a Complex World* (Prentice Hall: 2005), John Thackara provides an overwhelming argument for how design is influencing the shape of our world. With a world economy and global communication, the context for delivering education has to be a global one. This 'de-parochialization' includes considering the hidden cost of design on a global scale, and as teachers we need to encourage children to become aware of these global issues and trends to try to raise their awareness of the moral and ethical issues that are embedded in every product.

At current rates of product usage it is calculated that we will need three times the Earth's capacity simply to maintain current consumption levels. This is in the context of the world's population increasing by one million every 11 years.

When working in this context, pupils need to be learning of the implications of their decisions, not just as children in the D&T environment but also as future consumers and citizens. In doing so they can begin to challenge some of the misconceptions that exist about designers: that they are liable both to save the world as well as damage it through the actions they take.

By placing D&T in a global educational context we provide a satisfying stimulus for discussion and learning, and more importantly for informed action-taking behaviour by pupils.

Activity: Next time pupils are designing, get them to consider the global influence of their product. At the simplest level this might be to estimate the distance travelled by each of the component/ingredient parts from raw material to finished product.

The average power tool is used for ten minutes in its lifetime.

The average CD is used once.

Much of the design in the world is aimed at the wealthiest 10% of the population – however don't think that this 10% merely represents millionaires – this includes you!

The truth is that 90% of the world's population are not designed for. Of the world's total population of 6.5 billion, 5.8 billion have little or no access to most of the products and services that we use on a daily basis and take for granted. Almost half of this same group do not even have regular access to food, clean water or basic shelter.

It is important that pupils, through D&T, recognize how lucky they are, how their decisions impact upon others' lives and their responsibilities to the global community. This can be done through exploring designs for the needy rather than the wealthy, recognizing the exploitation of workers and recognizing the environmental damage that can occur through mining the earth's valuable resources. Humanitarian design has to be a key part of all pupils' D&T experiences.

Activity: Consider introducing activities where children are designing in contexts such as disaster situations, e.g. in famine-hit areas and developing nations.

GLOBAL EDUCATION (2)

There are undoubted links between creativity and innovation and both are at the heart of D&T. But what is the difference? Generally creativity precedes innovation in that it is the moment of inspiration, while innovation is the period of transformation. Therefore, in one way, innovation can be considered as the implementation of creativity, linking one to the other as part of a continuum.

Innovation is, however, critical to the creative D&T experience, and Professor Richard Kimbell's 'Assessment of Innovation Project' has neatly defined the means for both recognizing and potentially assessing innovation, through the recognition of a pupils ability to have ideas, grow ideas and importantly prove these ideas.

A key part of any innovation is the ability to reconcile tensions with the problems experienced, and this invariably means dealing with and engaging with others, possibly as part of a team. This type of activity has typically been missing in many D&T classrooms, but as exam boards begin to recognize that much of the coursework that takes place in D&T is often contrived, opportunities are increasing for pupils to be innovative and creative.

So answer the following questions and reflect on whether you are encouraging innovation:

o What percentage of activities do pupils work in groups?
o What percentage of activities encourage open-ended opportunities?

If the answers to these two questions are less than 40%, the balance may be wrong.

To recognize and achieve innovation in your classroom is really quite simple and straightforward, starting with three simple changes:

First, adopt the criteria for recognizing innovation and assess pupils purely on these, namely: having ideas, growing ideas and proving ideas.

Second, create a fruitful context for the innovation to take place. The length of time for the activity is down to personal choice and can be anything from a single lesson to a series of lessons.

Finally, a key part of this is encouraging teamwork and communities of practice. This can either be in the ideas stage, helping the ideas grow, or throughout the activity. Just by adopting these three simple steps you will begin to see big differences both in pupil motivation and creative and innovative outcomes.

INNOVATION (2)

One of the features of D&T that always surprises me is that we often ask pupils to suspend their understanding of, and participation in, the 'designed and made' world when they enter the D&T environment – whether that is a studio, kitchen, lab or workshop.

We ask them to lower their standards and to forget about the quality, or impact on the environment, of the products they make in the production of often poorly conceived, poor quality 'knick-knacks', 'snacks' and 'garments' which in the real world they would show no interest in at all.

Outside of school theirs is a world of fashion and quality products perpetuated by mass consumerism, while on the inside it is a world of 'shaky hand' games, scones, key fobs and slippers perpetuated by the myth of the need to 'take something home'.

At one level perhaps this suspension of their quality control systems represents an appreciation of their toils and troubles in reconciling and overcoming difficulties in the production of an object – which may in itself, if accompanied by sufficient learning, justify a limited rationale for such activities. However, the fact that teachers are willing to usher pupils into the suspension of these critiques in the production of such inappropriate items is a source of some concern.

The paradox of on one hand producing discerning and powerful consumers as part of the clear rationale for D&T, compared with the participation and acceptance of poorly conceived, superfluous product outcomes.

It really is important therefore that teachers ask themselves some fundamental questions before putting a child through an activity, such as:

○ Is there sufficient learning taking place?
○ To what extent does the end justify the means?
○ Who benefits and who loses by this activity?
○ Does it represent progression?
○ Does the activity require a suspension of values?

School culture has changed in the last few years, and whereas previously the notional head of department was seen as the person responsible for general sourcing, managing technicians and so on, their new role is very much characterized as being 'leaders of learning'.

This new focus puts teaching and learning very much at the heart of what the subject leader does, and as such a key part of this is being aware of the latest thinking related to teaching and learning, as well as monitoring the quality of teaching and learning within a group of teachers.

To be able to do this, leaders of learning have to be able to understand, demonstrate and deliver what high quality teaching and learning looks and feels like. This can be demonstrated in three areas:

o Knowledge – leaders must have knowledge 'breadth'.
o Pedagogic knowledge – leaders must have 'depth'.
o School knowledge – leaders must know how to apply 'depth' and 'breadth' in the context of their learning environment.

The essence of a good leader of learning is understanding the people working with you, but before that you have to understand yourself. A key part of this is having a clear philosophy of what you believe good teaching and learning in D&T looks like. If you have never done this, try and write it down in one sentence.

Remember that your sentence is about your philosophy of learning in D&T and should not focus only on what you want pupils to make. Once you have created your sentence, share it with your team and let it grow until you can come to agreement for locating learning in D&T. Once this is agreed all future developments can be considered in light of how effectively they contribute to the agreed statement of learning. If the developments do not contribute then you may feel they don't warrant inclusion.

Scenario: first lesson in D&T in Year 7. 'Right children – remember you visited here for an afternoon in Year 6 and did all that exciting stuff? Well, you won't be doing that again for another four years. And remember when in your last school you were able to make decisions about what you were doing, and were treated maturely as you were the oldest in the school? Well, now you're the youngest and lowest priority, so you can forget about making any decisions for at least three years. Now we are going to spend three weeks writing out the safety rules, doing a "spot the danger" worksheet and watching a stomach-churning safety video.'

Is this a familiar scenario? Does it contain a grain of truth even if exaggerated a little? The truth is the first seeds of dissatisfaction and disaffection are sown in those first few experiences of D&T in the secondary school; instead of capitalizing on pupils' energy and creativity, we often resort to giving them a deadly dull experience for three years and then suddenly ask them to design and be creative in Year 10 – is it any wonder that they then respond by saying they cannot design?

So, if you insist that they have to study safety in the first few weeks, perhaps this could be done through drama and role play? Could their first experience build on what they might have encountered on open day, and could they be the top priority instead of the lowest priority? Investing in them will pay dividends while neglecting them will cost you dearly!

Bungee jumping or bridge building?

How do you view the primary to secondary school transition for pupils? Some will take a bungee-jumping approach: 'Close your eyes, away you go and I hope you survive!' Others will take a bridge-building approach, where pupils are escorted across the primary-to-secondary divide in a strategically planned manner.

Clearly any discussion about primary school to secondary school transfer depends a lot upon the context in which the transfer is taking place. Some secondary schools will have two or three main feeder schools, while some schools will have 40 or more primary schools to deal with. The reality is that secondary schools are wasting an enormous opportunity if they do not tap into their new pupils' previous six years of D&T experiences, and the worst case scenario is where the secondary school assumes that the pupils have learned little that is useful in this subject. Do you know what your pupils are learning in their primary schools and vice-versa?

Previous research into transition has shown that children often take a dip in performance across the primary to secondary transition stage due to variances in teaching styles. This is largely due to pupils acting autonomously in the primary school while the teacher acts as a facilitator, whereas in the secondary school the pupils are merely told what to do, as if they had no previous experience, while the teacher acts as an instructor – ultimately limiting the pupils' learning capacity due to predetermined outcomes which have in-built limitations.

Activity: A key place to begin any discussion relating to pupil transfer is considering teaching, learning styles and pupil expectations. Given that pupils arriving in secondary school will have spent six years studying D&T in some form, it is worth investing sufficient time exploring this rather than adopting a 'clean slate' approach.

THINKING ABOUT TRANSITION (1)

THINKING ABOUT TRANSITION (2)

The ideal scenario in any secondary school is that each member of the D&T department has a linking role with at least one or two primary feeder schools. In this situation the process of transition becomes a two-way process and within this context it is critical that the secondary school is not merely seen as manipulating the primary school and trying to achieve the upper hand. The key is understanding the different ways in which the subject can be delivered within the different contexts and how the transfer can be managed successfully. The most common way for this is the transfer work scheme, where pupils begin an agreed activity in Year 6 (primary) and continue it into Year 7 (secondary).

In such situations it is not so much the scheme of work that overcomes the difficulties, but the outcomes of the rewarding discussions between the teachers in the primary and secondary school setting concerning the pupils' learning styles across the transition divide.

Some other tips:

o Consider your Year 7 experience as being more like the primary schools', rather than trying to get the primary school to be like the secondary school.

o Instead of trying to develop skills when considering transition, consider what vocabulary pupils will have in the primary school that will enable them to access the secondary school experience.

o Although a transition scheme of work may not be possible between all the different primary schools, see if pupils from each school can compile a 5 minute PowerPoint presentation of their last year of D&T experience which can be used for 'show and tell' in their first D&T lesson. This way each teacher has a much more informed starting position.

All curriculum subjects are expected to contribute to the SMSC entitlement of learners. D&T by its very subject matter is a rich activity for developing and exploring these issues, such as:

Social – considering how products engage or disengage users with other users. This might include considering how the MP3 player has meant people listen to music more and talk to each other less. Is there a way of designing people in to, rather than out of, social interaction?

Moral – morality is clearly closely linked with values and all decisions that learners make are value laden. Values are translated into products and children need to consider their values in the decision-making process. For example, what are the moral values in using material and products from impoverished parts of the world where child labour is commonly used? What are the moral values in an overly consumerist society?

Spiritual – by taking a broad approach to spirituality, products can be considered in terms of their ability to create richness in life. Such products can be uplifting and life-enhancing. D&T is also about improving the quality of life, which in itself provides valuable opportunities for the discussion of spirituality.

Cultural – D&T is about examining and enriching our culture. This may be through the examination of existing cultures or the design of products, systems and environments to engage with and shape future cultures.

Some possible broad areas of discussion: Economy/e-commerce; Environmental/green issues/sustainability; Conservation; Global issues; Cloning/genetic engineering; The food industry; Fashion and cultural icons; Science/medical technology; The digital divide; Globalization; The media; Peace and security; Politics and diplomacy; Transgenerational issues; Disability; Human rights; Transport; Sustainability.

SOCIAL, MORAL, SPIRITUAL AND CULTURAL (SMSC)

FROM QUALITY CONTROL TO QUALITY ASSURANCE

I once visited an aircraft engine factory and when speaking to one of the staff I rhetorically suggested that they must have a really good quality control system. He replied that there was no quality control process at the factory, as what tends to happen is people can get casual in their work if they know there is a safety net at the end – which you don't want when building aircraft engines! Therefore they employed very good quality-assurance procedures that were built into each stage of the process, rather than a final check at the end.

In many ways, teaching and learning is moving in the same direction, and Ofsted is an example of this way of thinking. Instead of waiting every four or five years to quality control a school's performance, they now expect schools to quality assure their own processes, providing evidence through self-reflection. Evidence from Ofsted suggests, however, that self-evaluation in D&T is not good and that evidence is often lacking.

Attempting to improve quality assurance in teaching and learning can be done in many effective ways, however one simple method is to focus upon teaching, and in particular peer observation and development.

Firstly an agreed theme might develop from a departmental meeting. This could be, for example, to focus on the quality of questioning in lessons. The follow-up meeting would generate discussion perhaps based upon literature on questioning that would be presented at the meeting. Following this, an agreed observation schedule could be drawn up for teachers to observe each other (more evidence) with a proforma for recording the nature, type and frequency of questioning being developed (further evidence). Peer observations would then generate additional evidence which would be used to generate discussion at the next meeting, followed by further observations and discussion (all generating extra evidence). Ultimately this activity would naturally be generating evidence as well as improving the quality of teaching through the improved questioning technique of teachers. The data provided from agendas, minutes, proformas and observations would all provide excellent quality-assurance evidence.

To what extent should the individual fit the system, or should the system fit the individual?

In many ways, personalized learning has represented a term seeking a definition since it was first used by a government minister for education in 2003. Regardless of it being an embryonic and emerging term, it appears to offer teachers a means for reconceptualizing their role in educating pupils. For D&T teachers how this might look is still not clear, but it is not difficult to imagine how the six key themes below can be embedded into the subject:

○ Responsibility for learning
○ Confidence as learners
○ Engagement with learning
○ Maturity in relationships
○ Independence and control of learning
○ Co-constructors with teachers in learning.

Essentially, good D&T offers all of the above. If pupils are given genuine opportunities to design and take action then the rest should follow. Many advocates of D&T would say that this is why the subject is unique, as personalized learning is naturally embedded in the subject. What may change however is when, where and with whom personalized learning is developed, which provides even more interesting opportunities.

The concepts of vertical grouping – learners being grouped by stage not age is one potential change, as is e-learning, peer-to-peer work, etc. What it all amounts to is a customization of education and a system where individual differences are not magnified by the system but can instead be accommodated. Such changes have already been seen in many professions, most notably the medical profession, where a whole range of paraprofessionals work alongside each other. In a maturing education system it is anticipated that such diversity will be forthcoming in schools, at the heart of which lies the addressing of individual needs.

Activity: Consider the list above. How might these be enacted within your teaching?

'We can tell our values by looking at our checkbook stubs.'

Gloria Steinem

Values play an enormous part in D&T and it is critical that teachers firstly understand and question their own values, and secondly get pupils to do the same.

There are many opportunities in a lesson for rewarding discussions about values, as every design activity involves pupils translating their value judgements into an outcome of some form. At the simplest level this may be a choice of colour, while at a more complex level this might involve questions about the appropriateness of introducing new products which consumers may not need.

A simple starting point, however, is to ask pupils what they value and why. From a list of values pupils can begin to explore if these values are shared within a family, a small community (a school or village), nationally or globally. The critical path is to examine other people's perspectives and see how different value sets interact and can occasionally cause conflict.

Such values may be located spiritually, socially, economically, morally, environmentally or technologically. Regardless of the location of such values, it is important pupils can both defend their own values and the transmission of these value judgements through the decisions that they make when creating outcomes in the form of products that others will interact with.

For this type of activity to be really successful, good contexts for designing have to be available for pupils, as if the value decision-making activity is removed, such as a project where pupils have little ownership, then pupils will be denied a worthwhile learning opportunity.

It is very tempting to consider the Key Stages as one coherent phase with the main focus being to complete the relevant Key Stage. In doing this a trick is missed as each year has the potential to effectively be chunked to provide specific focus and orientation across the year. Such a focus helps to contextualize the year within the broader context of the Key Stage.

A starting point is the statement of importance. By taking elements of this and dividing it up both across each year and within each year gives a much sharper focus to the activities. Providing an overall orientation in each year helps get the maximum from each Key Stage.

For instance, a way of conceiving of Key Stage 3 may be that Year 9 is going to be delivered in four chunks, with each one focusing on an element from the statement of importance, e.g. cultural, economic, emotional and historical. You then may decide that the overall context for the year is D&T and careers. The reason for this is that pupils probably will be making career choices in this year.

Therefore activity one in Year 9, as well as delivering elements of the programmes of study, will link both culture and careers in relation to D&T.

Thus across the Key Stage, assuming a context-led approach (although there are alternatives), we could deliver learning as per the table below, with the key being that the broad contexts provide the orientation to the additional richer learning:

	Context 1	Context 2	Context 3	Context 4
Year 7 focus: People	Aesthetics	Technical	Social – Working as teams	Industrial
Year 8 focus: Industry	Innovation	Social – Working as teams	Past and present	Products and systems
Year 9 focus: Careers	Culture	Economic	Emotional	Past and present

Designing

Designing is considered to be changing one set of circumstances for another – but any intelligent fool can make things bigger and more complex . . .

STIMULATING CONTEXTS (1)

If there is a recurring theme within this book it is the concept of delivering D&T through rewarding and stimulating contexts. Good contexts enrich activities by engaging and motivating pupils as well as challenging their emotions and values. Poor and barren contexts disengage and de-motivate pupils.

The setting of contexts is not however a simple task and in the same way as knowledge, skills and concepts operate along a progressive scale (Vygotsky's Zone of Proximal Development is one way of considering this), then so should the contexts. This means that the context through which a project is delivered shouldn't be one that a pupil is totally familiar with at the start of the activity. The ideal context is one where a pupil can locate some understanding and has some attachment to it, but by the end of the activity they will have become immersed and knowledgeable about that context.

When teachers are choosing contexts for pupils to work in they shouldn't pander to the pupil's immediate interests as often this only draws upon the learners' existing knowledge, values and prejudice within that setting. It is only through taking pupils on a journey from the ill-conceived to the conceived that the experience becomes a valuable one.

Such contexts could include looking at social phenomena, medicine, new technologies, disabilities, cultural differences or environmental and sustainable issues, and could be developed in any of the material areas of the subject.

An essential part of this is to 'hook' the pupils by stimulating them with interesting, provocative questions and information about that context. From this a whole range of discussions can take place around misconceptions and understanding before revealing the context of the D&T task they will undertake.

Often a reason for not providing a rich context is the fear that pupils cannot make what they design. To counteract this argument, it is worth considering some of the examples from the Young Foresight project (www.youngforesight.org) where pupils have been involved in challenging 'design without make' activities which have been set within fulfilling learning, engaging and empowering contexts.

STIMULATING CONTEXTS (2)

EVALUATING

The prominence of evaluation within D&T over the years has varied. In the earliest version of the National Curriculum evaluation was one of the four attainment targets and carried significant weight; however, overall it is something that has often been done poorly, often consigned to the end of the project as a time filler.

Yet evaluation offers the opportunity for genuine higher-level thinking and if you are familiar with Blooms Taxonomy, shown below, then you will recognize that evaluation is considered the highest level of cognition.

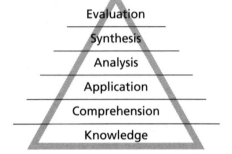

Evaluation
Synthesis
Analysis
Application
Comprehension
Knowledge

Bloom's Taxonomy

As evaluation is considered a high-level activity, its role should be much more prominent within the subject. This means that evaluation could initially be used to signal the start of an activity rather than the end of one. By evaluating systems, artefacts and environments at the start of the process we can use this critique to generate design opportunities and responses. For successful evaluation it is important that the student engages in the appraisal process, assesses on the basis of specific standards and criteria, and that they are able to judge, recommend and justify their decision making.

Evaluation is naturally clearly linked to criteria, and generation of such criteria, by which pupils can judge and critique their own work and that of others throughout the product development, is a critical part of the evaluation process.

Part of the National Curriculum's statement of importance mentions pupils learning about improving the quality of life. This is a profound ambition and I would contend that before this can be achieved whatever we do in D&T must not make life worse than it already is.

This might sound a glib comment but before any activity is undertaken in the subject our pupils, as consumers, will have already consumed an enormous amount of materials. In their short lives they will have already consumed as much, if not more, materials and energy than their grandparents, and certainly their great-grandparents.

At current rates of consumption it is calculated that we will need three times the earth's available resources to maintain our current consumption rates. This is compounded by our belief that we 'need' more – which promotes even greater consumption. For instance, John Thackara tells us it takes 1.7 kg of material to make a 32 MB RAM microchip, that is, 630 times the mass of the final product (*In the Bubble*, Prentice Hall: 2005). The amount of waste matter in the manufacture of a single laptop computer is close to 4,000 times its weight – 15 to 19 tons of energy.

To begin to design more unwanted items in D&T without critiques of the environmental and sustainable implications would appear foolish, and to not critique current consumption in the context of sustainable, environmental and global issues would seem reckless. A good place to begin discussions with pupils related to this can be found at Practical Action (www.itdg.org).

THINKING ABOUT SUSTAINABILITY (1)

THINKING ABOUT SUSTAINABILITY (2)

Although the quantity of materials used in D&T through designing and making are considerable by themselves and remain a concern, it is perhaps the hidden message – that the consumption and production of unwanted and unneeded items is acceptable – is the bigger concern.

A simple example of how D&T can look differently at this issue is exemplified by packaging. Virtually every school I know delivers a packaging project, yet often the brief involves making more rather than less packaging.

However, by taking the concept of an existing product, critiquing it and reducing the amount of packaging, and then taking further action by contacting the company responsible for the excessive packaging, seems to embody the key principles of good D&T. Fortunately, there is lots of advice about designing in sustainable ways such as Sustainable Technology Education Project (www.stepin.org).

*'My kids just can't design – you ask them to design
something and they just get stuck on one idea!'*

Do you recognize the scenario where pupils get fixated
on one design and find it very difficult to change
direction? Part of the problem lies in our own
expectations and the culture of schooling. In most
subjects, five lessons a day and five days a week, children
are told what to do and when to do it: 'Name on the left
hand side in pen with two lines immediately
underneath'. Suddenly they are asked to come up with
new ideas and often they fearfully stagnate and stick with
the tried and trusted ideas.

To break this cycle it is really important that the
teacher uses a range of strategies that prevent pupils
from being fearful, and encourage and reassure them that
it is good to experiment and make mistakes – however, if
the teachers say this they have to mean it.

Getting children to work in this way often means
changing the dynamic – so instead of working on their
own with white paper and pencil at a desk, why not get
them designing in groups on the floor in pen on brown
paper, or get them modelling in Play-Doh or foam.
Using different strategies breaks the fixation, so using
different idea-generation methods such as 4 x 4, walking
on the wild side, 50 circles, etc. will all help.

Finally laugh, praise, share and most of all reassure
pupils that what they are doing is good!

PLAY

One of the most common methods of encouraging creativity is through play, yet play may often seem to be the antithesis of what should go on in the D&T environment. But without giving children time to be playful with their ideas, the chances of genuine creative outcomes are low.

How do you construct a playful learning environment? Well, play doesn't mean unstructured, do-what-you-want type chaos. There has to be a set of ground rules as in any activity. However, the opportunity to be risky and to play around with ideas without fear of getting the wrong answer has to be encouraged as part of this.

Such freedom for pupils within lesson time is not usual, and within the classroom you may have to model the very traits that you want to encourage – it is important pupils see that not knowing how to proceed is an acceptable form of learning. The teacher who says 'I don't know, but we can find out together' is modelling good practice (as long as it isn't something they really should know!).

In fact not knowing how to proceed, and using a variety of other skills such as negotiation, research, planning, testing, questioning and social skills, are the very skills that we have to encourage in children to enable them to become autonomous decision makers.

One tip for encouraging this type of behaviour, one that is used in many creative agencies, is the brandishing of a yellow card for anyone who, when generating ideas, is being too predictable, boring or is being too negative with others' ideas. Therefore, the teacher or pupils can brandish the yellow card whenever it is playful idea time and someone (including the teacher) isn't being playful enough.

In many classrooms children are asked to sketch their ideas before they can proceed to the next stage. This often meets resistance with pupils who often feel they don't need to sketch, or who feel they are not very good at sketching. There are two important points that are needed to be explained to pupils to help overcome these misconceptions.

Firstly, why do we need to sketch? Ideas tend to last in our short-term memory for a very brief time – often a few seconds. Have you ever had a brilliant idea in the middle of the night and forgotten it in the morning? This would be an example of the idea existing in the short-term memory and not being transferred to the long-term memory in some way: use it or lose it! The reality is that unless something is done to transfer an idea from the short-term memory to the long-term memory, where ideas can then begin to be manipulated, then the idea is likely to be lost.

When designing we often record ideas in the form of a sketch, as it is quick and it turns the idea into a visual stimulus which helps the transfer of information into the long-term memory. Quick models (e.g. made from Plasticine, card, etc.) can also work in the same way, and once the idea is externalized through modelling or sketching it can then be re-internalized through visualizing the sketch, which aides further manipulation of the idea.

A second misconception is that sketching has to be neat. The reality is that rapid sketching is likely to be messy, and therefore pupils should be encouraged to sketch rapidly without fear of it being commented on in relation to neatness or accuracy. Sketching in itself serves different purposes. A sketch for developing ideas can, and perhaps inevitably will, be rough and intuitive. As such, it merely serves the purpose outlined above. Representational and developmental sketches can be used to aid communication with others and, therefore, will need to be more accurate.

THINKING ABOUT THE BRAIN
WHEN DESIGNING (1)

The brain is a wonderful organ. It starts working the moment you get up in the morning and does not stop until you get into school!

It seems pretty absurd but for a long time the brain was largely ignored when thinking about education in schools. However, in recent years, many schools have invested heavily in an attempt to develop more brain-friendly learning environments. These have included using music in lessons and corridors, changing room layouts, providing water for children, Brain Gym™ and so on. The extent to which these ideas are effective is uncertain, however, it does indicate recognition that some understanding of the brain and how it learns may be helpful.

The brain mechanics of sketching have been dealt with previously in this book, but another key area relates to how the brain deals with uncertainty, anxiety and risk. Ultimately the brain is built for survival and it will always resort to this default state when there is a sense of fear or anxiety. Unfortunately, the conditions that we operate under when studying design can be very close to these areas. Often we have to work in very uncertain and risky ways, creating a chemical called acetylcholine in the brain – which creates that unpleasant feeling of not knowing how to proceed. In such situations the amygdale which is responsible for the fight-or-flight syndrome can override the thinking part of the brain, resulting in a lack of creative response as it struggles to regain a sense of certainty.

Therefore, a key part of any designing activity is the reassurance of pupils through the teacher's use of language. Try also to exemplify good practice in showing how it is okay to not know how to proceed with an idea – for instance, showing how designers often struggle – and explain that uncertainty is an essential part of designing.

Within the learning and designing environment there has to be a strong sense of trust between the teacher and pupils, and this sense of trust naturally creates a chemical in the brain known as oxytocin. The stronger the indicator of trust, the more the levels of oxytocin increase, and when this is observed by others trust can increase throughout members of a group – so your group will tend to be all with you or all against you. What's more, if from this sense of trust comes a sense of success and expectation then the brain creates dopamine as part of its reward system, making you feel good and making you want to revisit the feeling.

Therefore, if the teacher creates an environment where there is low threat, high challenge, trust and success, this sets off a whole series of chemical reactions in the brain which are much more likely to result in overall success in the student. Equally, high anxiety, stress and a lack of success can have the opposite effect. To create trust and high expectations consider how you use language and use statements (if true) such as:

o 'I know you are all capable of coming up with really good ideas'
o 'There are some really good ideas coming out and you are showing real signs of improvement'
o 'There was some fantastic work today and I look forward to next lesson to see how your ideas have developed'.

THINKING ABOUT THE BRAIN WHEN DESIGNING (2)

RECOGNIZING EMOTION – PERSON

'"Come to the edge," he said. They said: "We are afraid." "Come to the edge," he said. They came. He pushed them and they flew.'

Guillaume Apollinaire

We are all driven by our emotions – they provide our ordinance system and drive us and motivate us. Our emotions steer us also as teachers and they steer the pupils that we teach. Given that emotions are such a large part of what we do, perhaps we should consider how our teaching influences our learners' emotions?

From my own research I have divided the location of emotion into three domains: Person, Process and Product. This is aimed at encouraging and promoting a genuinely creative, learning and product-orientated experience.

Within the first stage, Person, we need to consider how we encourage creativity and how we engage with pupils. Therefore when encouraging pupils to design and be creative, do you give out life jackets or straightjackets?

In other words do you – through the language that you use, the body language that you adopt, the facial expressions you have, the exemplar work that you show and the tasks that you deliver – encourage pupils to take risks while giving them the confidence to enjoy being uncertain and not knowing how to proceed? The reality is without this it is unlikely that your pupils can begin to take risks and be creative.

Teacher: 'What is it with you – ignorance or apathy?'
Pupil: 'I don't know – I'm not really bothered!'

IDEA
36

Building upon recognizing the location of emotion in the Person stage, the Process stage represents the emotions involved in the process of learning. This stage links fluidly with both the Person stage and the next stage (Product), and is concerned with the learners' engagement and motivation through emotionally engaging and challenging contexts.

Within this stage it is useful to understand how the brain works. Firstly, our feelings and emotions control our thinking and not the other way around. Therefore, we are trying to engage the learners' emotions to facilitate thinking. Failure to emotionally engage can result in our emotions rejecting the activity.

Secondly, behavioural neuroscientists have also suggested that the naturally-occurring chemical, dopamine, within the brain mediates some aspect of reward learning, or the capacity to predict rewarding events. Therefore, success is attachable and events such as teacher praise can generate positive emotions in the learner.

Dopamine release in the brain also switches on our attention system and facilitates thinking in the frontal brain, before facilitating cognitive engagement leading to learning. This type of learning has been called 'flow', the moment when you are completely immersed in your work.

Finally, within the learning phase is the concept of implicit learning (learning in which you don't know that you're learning and when you don't know what you've learned). This is very much what happens in D&T when learners become immersed in an activity and are emotionally attached and committed to seeing the activity through. Such learning is very powerful as long as pupils are given an opportunity to reflect upon the learning process.

Activity: Consider how you engage pupils emotionally in the process stage of emotional learning. Think about whether the contexts that you use are emotionally engaging, stimulating and challenging or are they tokenistic and contrived?

RECOGNIZING EMOTION – PROCESS

41

The final stage of the location of emotion in D&T is in the Product stage. In this stage we are concerned with two areas of emotion.

Firstly, we are considering which emotions are concerned, and how these emotions will impact up the users of the designs that we create. The second area is educating learners about how their emotions are manipulated by the products, services and environments that they come into contact with.

Therefore, if a pupil is designing a product – whether it is in a food, graphic, electronic, textile or material context – it is important to consider how they want the user to engage with that product and what emotions and feelings it should create. The easiest way to do this is both through language and visual mood boards.

In the Product stage it is also important that there are links with the first two stages, Person and Process, so that the learner feels as if they can take risks with their ideas and that they are engaged through a stimulating context.

Some emotions and 'feeling' type words that you may want to create in the user that you might consider when designing are:

Alienated	Bashful	Cautious	Determined
Ecstatic	Fearful	Guilty	Happy
Interested	Jubilant	Kind	Longing
Mischievous	Negative	Optimistic	Petrified
Quizzical	Relieved	Sad	Thoughtful
Undecided	Vindicated	Withdrawn	Xenophobic
Young			

Remember – tight briefs restrict thinking!

PROJECT SEMANTICS

We all know what a bin man does but if we change the language and call him a refuse technician we suddenly have a different perspective on the role. Although it might include 'bin men', it broadens up the role to consider lots of other possibilities, possibly even exciting opportunities, which such a person may be involved with.

The same is true with D&T projects. For some reason, when you ask a teacher about what the curriculum with a particular year (e.g. Year 8) is, they will say something like 'We are doing the pencil case/slippers/scones/shaky hand game project' and so on. What this reveals are two key issues at the heart of D&T: teaching and learning.

Firstly, the emphasis in the answer is on doing. The answer is not usually 'We are learning about material stresses/joining materials/etc.', but that we are simply doing something which suggests a means to an end rather than a means to learning.

Secondly, whatever the pupils design, regardless of how brilliant (or not) it is, it will always end up as a predetermined solution governed by the title – which although it may allow for some minor embellishment, will ultimately be constraining. At times, there is clearly a legitimacy in having focused tasks, but equally there have to be opportunities for pupils to explore ideas from a less constrained starting point.

Considering the semantics, instead of labelling the project with the solution/result, why not describe the context of the problem? Therefore, pupils will be learning in a desk space context/recreational foot protection context/funky food context/electronic fun context. By pulling back from the solution to reveal the bigger but more blurred picture, we provide real opportunities for pupils to think and explore their ideas without the constraints of a highly-focused picture posted in their 'minds eye' by the project title.

WHAT IS COGNITIVE MODELLING?

For some, cognitive modelling lies at the heart of the D&T experience, but what it actually is might not be so clear.

One way of viewing it is to imagine a child playing with a Meccano construction kit. Think about the moment when they have to fix a nut to the inside of something and realize the only way is to wet the end of their finger to stick the nut to it. Visualize what the inside of the space might be like as they manoeuvre the nut closer to the bolt. Visualizing is seeing something that cannot be seen, and is a high-level skill – in essence it is 'imagineering'.

In summary, this is the meaning of cognitive modelling. It is the images that we form in our mind where we visualize what we cannot see, in our 'minds eye'. The ability to manipulate ideas in the mind, to visualize and manoeuvre them is a high-level cognitive activity, and is a critical part of D&T. However, such modelling does have limitations, and the consecutive stage of externalizing the ideas is an important part of the further refinement of ideation.

This is known as an iterative process and it is central to the teaching of D&T. It represents the interaction between the hand, the eye and the mind, and the simplest way of encouraging this is getting pupils to visualize the inside of products, for example, considering the internal workings of a product and asking students thought-provoking questions by getting them to imagine what is happening inside, and why?

At times within this book I have referred to design without making (or manufacture). This is not that I have anything against making, but from a learning perspective most learning takes place in the early stages of the designing and reconciling processes, whereas most consolidating takes place in the manufacture stage, particularly where the manufacturing is predetermined.

A further weakness of a preoccupation with manufacturing is that it can inhibit designer-type thinking – if pupils are constantly thinking 'this is a good idea, but I can't make it' then they are less likely to be ambitious or creative. It is equally important, however, that pupils do have an understanding of material proprieties and processes.

Concept modelling offers a partial resolution to this as it allows the conceptualizing of an idea through the production of a representative model. This is an entirely legitimate activity and most pupils will already be familiar with the context of concept houses, cars and phones.

Through concept modelling pupils are able to more fully explore form, colour, ergonomics, emotions, shape, surface detail, weight and balance without having to fulfil all the requirements of the product development. So while they can make a concept model with the correct shape and form (e.g. in textiles or materials), elsewhere they might model the inner workings, such as circuit design, fastening, joining systems, etc.

Such practice is highly legitimate as long as it is not used as a shortcut for overcoming the problems of manufacture. Therefore, it is essential that concept modelling is used as a proving tool, and the development of the concept is designed to show that the intended design is achievable.

CONCEPT MODELLING

USING RESEARCH (FOR PUPILS)

If designing a chair and looking for inspiration, should you look in a book on chairs or a book on biology?

Most people will be familiar with this story: In 1948, George de Mestral, a Swiss mountaineer, went for a hike with his dog, and when he returned home both he and the dog were covered with burs from a plant. He looked at the burs under a microscope and found them covered with stiff hooks. Working with a weaver from a textile mill in France, de Mestral perfected the hook-and-loop fastener which he named Velcro and which was finally patented in 1955.

The chances are that the first thing a pupil will do when researching a project is Google the name of what they are trying to design. This is probably the worst thing they can do if they want to genuinely come up with a new idea, but the easiest thing to do if they want to come up with a similar idea to an already existing product.

When designing, pupils need to be encouraged to go back to first principles by exploring the problem and not possible solutions. By looking at the problems from different perspectives, we can give a whole new outlook to an existing problem. Therefore, the sources of information that learners use to influence their design is really important – for example, taking stimulus from the natural and created world can give pupils much more added impetus than simply looking at other peoples' solutions.

Have a look on the walls and bookshelves in your classroom to see what is displayed and, if there are only pictures of existing products in the room, then you may be missing an opportunity.

The answer to the question at the top of the page is therefore that you could look in a biology book and examine both the natural form of the body as well as the natural world – how do other animals rest, what are the best ways of supporting a body and what natural forms of beauty can we use to inspire our designing?

It takes a touch of genius – and a lot of courage – to move in the opposite direction.

Albert Einstein

○ Are you creative?
○ Are your pupils creative?
○ Does your teaching encourage creativity?
○ Do your pupils' products show a creative response?
○ How do you know?

One of the great features of D&T is that it is considered to be a creative subject, in fact it is the only subject on the curriculum where creativity is mentioned twice in the National Curriculum statement of importance.

However, for many, creativity has been considered 'in crisis' within D&T as often teachers have understandably avoided creativity in preference for safer forms of performance. Regardless of this, many still insist that they are encouraging pupils' creativity: a good indicator that you are doing this is to estimate the number of opportunities for something to go wrong in a project – not just in the manufacturing or time management, but in pupils' ownership and direction of the activity. If the pupil ownership is risk free and limited to formulaic embellishment, as in the choice of colour or limited control of shape, then this is not likely to be a creative act.

If, however, pupils are engaged in genuine risk-taking (but not recklessness) in pursuing their ideas, and pushing the boundaries of their understanding in pursuit of novel solutions, then it can be assumed that they are being creative.

CREATIVITY IN D&T (1)

CREATIVITY IN D&T (2)

If you are going to think outside the box, then you need to know the inside of the box intimately . . .

Most of the creativity that takes place in D&T is known as little 'c' creativity. Some ways of encouraging creativity include:

o Providing space for playful thinking that is free from immediate criticism and discouragement.
o Encouraging pupil self-expression and not limiting to simple choices.
o Using different methods for encouraging creativity (4×4, SCAMPER, Walk on the wild side).
o Being aware of the differing contexts for the development of ideas, the role of intuition and subconscious mental processes (the concept that what we think is a good new idea might be because we saw it previously without realizing).
o Encouraging and stimulating free play with ideas, the use of imagination, originality, curiosity, questioning and free choice.

Remember, if you were brought up in a grey town in a grey house with grey friends then you are likely to create something grey. Exposure to as many different ideas gives children a worthwhile platform for learners to build their ideas upon.

If you suggested to a pupil who is struggling to brainstorm that they were capable of finding more than 100,000 combinations of ideas, they probably wouldn't believe you. However, by using a simple morphological analysis technique, pupils can quickly generate lots of possibilities.

For example, imagine pupils are redesigning a concept MP3 player – by using combinations of words alone you can see that we can generate a lot of possible ideas (as shown in the table below). By arranging groups of words, every route through the options provides one of the 236,196 possible combinations.

So, for instance, one route across the table when generating ideas might be a futuristic, organic, polished aluminium, metallic, sensuous high-end MP3 product. A further 236,195 combinations remain!

MORPHOLOGY

Style	Shape	Materials	Finish	Emotion	Cost
Space	Organic	Injection moulded	White	Sensuous	Low end
Funky	Thin	Rubber	Red	Precious	Mid-market
Futuristic	Thick	Wood	Silver	Fragile	High end
Retro	Chunky	Brushed aluminium	Gold	Hot	Exclusive
Pop	Rectilinear	Aluminium	Blue	Exclusive	
Simple	Bevelled	Leather	Polished	Angry	
Clean	Pebble	Combination	Dull	Sexy	
Childlike	Sharp	Chrome	Metallic	Cold	
Minimal	Fluid	Cloth	Spotted	Happy	
9	9	9	9	9	4
9	81	729	6561	59049	236,196

TRANSGENERATIONAL DESIGN

There are more people alive today between 70 and 80 years of age than all the 70–80 years olds that have ever lived! As such, their lifestyles and needs are emerging and changing the way we think of this age group. The new 'grey market' is one that is active, open to experiences and often with disposable incomes.

Transgenerational design is concerned with inclusive design which appeals to people of all ages and abilities. Pupils in D&T should consider these new and emerging markets, not only as a source for creating new opportunities for design, but also as working in this area provides a rich context for challenging stereotypes and misconceptions that they might have about the needs of different groups. Whether working in a food, materials, textiles, electronics or graphics context, there are real opportunities for reconsidering how we conceive growing older.

The reality is that the pupils we now teach will live even longer, but will inhabit a very different world. They will, however, contribute to the future material and designed world and they will also contribute to the attitudes that shape future societies. Therefore, issues over diet, entertainment, transport, fashion and health will all provide stimulating contexts for designing and discussion. What will our pupils that we teach now be eating, wearing or using in 2070?

'Good things come in small packages.'

Designing packaging is one of the most common activities in D&T, but generally this is done poorly. The basic error is that learners are encouraged to design more packaging rather than trying to design less or different forms of packaging.

Packaging design promotes such rich opportunities and ties in many of the areas already discussed including concept modelling, cognitive modelling, SMSC, emotion, etc. When delivering a packaging project it is really important to question the fundamentals of packaging in addition to the new opportunities for packaging, re-branding and re-conceiving the product. This means considering whether the product is needed in the existing format, or could it be combined, substituted, reduced or eliminated (SCAMPER is a useful way to consider this). As such the problem may not lie in the packaging but in the source, the product or transporting of the product.

If the product is to remain then it is important to consider sustainability issues including reducing material and encouraging recycling. Other opportunities for considering sustainable factors include investing more in the package (so that it can be reused), using organic packaging, or considering offset packaging as was done by Varta (batteries), who included plant seeds with every purchase.

When considering packaging it is also important to consider both the ethical dimension (encouraging people to buy products they don't need) and the cultural impact of the product being packaged.

When delivering an activity such as packaging it is also critical that we remember the reason for delivering such an activity. The first premise is not that we are trying to make future packaging designers but that we are trying to educate and empower future consumers who need to be discerning and informed in the choices they will make.

PACKAGING

THINKING ABOUT DESIGNERS – PROS AND CONS

It is tempting to think of designers in a fixed way, and that is the 'celebrity' creator who adds value to a product. Such thinking has very much been part of D&T, as often we offer children the opportunity to become educated about designers based upon their D&T experience. However, as with all activities, we need a sense of balance – although designers do create wealth and employment there is often a hidden cost. This may be through the exploitation of workers in the production of the designed products, or in the 'design for obsolescence' approach – the deliberate creation of products which have a limited shelf life, which can have a significant negative environmental impact.

Victor Papanak, a great writer on design, writes about designers as both good and evil. On one hand they create products that can literally save our lives, or at least improve the quality of our lives through the products that we engage with. He also considers how designers are also responsible for significant environmental damage – such as in excess packaging or products we don't need; death – as in creating faster cars; or unhealthiness – as in encouraging certain lifestyle options such as drinking, smoking or obesity.

A good idea for the classroom is to consider the pros and cons of everyday products, as it is important that children recognize that design is not a one-way street: the values they have are a critical part of the role of the designer. It is these values that are at the heart of design and are something that should be given sufficient prominence in our activities.

Using technology

'*It has become appallingly obvious that our technology has exceeded our humanity.*'

Albert Einstein

USING CAD/CAM

The increase of Computer Aided Design (CAD) in schools has been remarkable, but a possible consequence is that children's activities have become even more constrained. Children now have unprecedented access to new technology, which should open up new ways of realizing their creative capabilities. However, the reality is that sophisticated and expensive machines are often producing objects within minor variations of a theme which children then embellish (e.g. put their initials on), rather than providing empowering opportunities for children to have increased ownership and to be motivated by allowing them to speculate on their own ideas. A misconception of CAD/CAM is that it gives validity to bad practice, just because it utilizes new technology!

CAD/CAM does, however, provide real opportunities for increased creativity, but what is needed are new models of teaching and learning within D&T (as opposed to adoption of industrial models of manufacturing). CAD/CAM has developed from an industrial desire to make products quicker, not an educational desire to increase learning. Producing products faster is not always desirable in an educational environment, as time reflecting and incubating ideas is equally as important (if not more so) as time spent manufacturing. Therefore if CAD/CAM facilitates faster manufacture and removes some of the drudgery from the process, more time should be available at the initial stages for the employment of creative methodologies, incubation of ideas and visualization of concepts.

The unfortunate consequence of the inappropriate implementation of CAD/CAM is that the amount of high-quality products produced in schools is often proportionate to a reduction of pupil ownership and engagement in the task. Therefore, ask yourself is CAD/CAM constraining rather than liberating ideas? If so, you know what to do!

USING WIKIS

A wiki is a website owned by a community which allows the users to control, populate and change the content. The most common example is Wikipedia which is an online encyclopedia constantly updated by its members.

Within D&T wikis can be really useful for pupils to work collaboratively on a particular project. For instance this might be through the teacher using the wiki to provide the brief or context for the activity to take place. The advantage of the wiki is that it is web based, so the stimulus can be much richer, such as including web links and video. From there pupils can be responsible for posting their research onto the wiki as well as providing a forum for discussion of some of the issues.

Using the wiki means everything is shared and this means that all research is shared across the group, with parents and even with other countries, for example, when doing a collaborative project with another school. As the project develops the wiki grows, and can become a lasting resource for others to build upon, or it can be merely wiped clean and started again – both have their merits.

The wiki can also be used as a means of assessment – instead of a standard folder submission at the end of the activity, the submission can be in the form of digital content such as photographs, voice files and video, giving a much stronger feel for the developmental work than traditional folders.

The key to the wiki is that it grows through ownership, sharing and discernment, and the skills and attitudes learned through this are very much those that are needed to operate in today's society.

IDEA 50

USING E-PORTFOLIOS

Did you enter the teaching profession because you were good at assessing?

The answer is likely to be no, yet a large and critical part of teachers lives involves assessing pupils' capabilities. Such assessments are often particularly difficult as we are often dealing with qualitative issues regarding how good something is, or how effective something might be. Even more difficult is when trying to assign a grade or a number to these areas related to creative or innovative ideas.

In many ways this is a square peg in a round hole scenario – you are obliged to do assessments of this kind, but the assessment method just doesn't seem to fit what you are trying to assess, and traditional means of assessment of pupils' work just don't simply seem to work. This often results in contrived outcomes in the form of 'jumping through hoops' through the tyranny of the portfolio.

Digital portfolios or e-portfolios offer an opportunity to overcome some of these difficulties. Firstly, a digital or e-portfolio is a form of multimedia storage and organizing system. At its simplest this may simply be putting everything on CD, a wiki or other type of web space. By doing this, it immediately means that pupils' assessment can be done in a much more effective way as pupils are able to include digital and animated images or sound and video files meaning that a much richer and more intuitive form of submission can be assessed.

Even more exciting is work taking place at Goldsmiths College along with the QCA which will ultimately allow online assessments of pupils' portfolios. Although in its infancy, the potential for online comparative assessments of pupils' work using digital technologies means a much more informed and democratic form of assessment and the possibilities for recognizing the diversity of capability will finally be possible.

Activity: Consider how electronic portfolios might enhance your pupils' D&T experience. This might at first mean considering submission of pupils' work on CD and move towards more sophisticated online assessments of pupils.

Extended curriculum

'It is easier to change the location of a cemetery, than to change the school curriculum.'

Woodrow T. Wilson

ENCOURAGING LITERACY

D&T is such a valuable subject for developing literacy. We often forget the vast range of different literacy skills that we teach children in the course of a lesson and often think our only contribution is through the development of technical language. Although important, this represents just one facet of a large range of skills and concepts that can be developed through D&T activities. For instance, explaining, evaluating, convincing an audience or client, communicating ideas, using technical vocabulary, comprehension and synthesis, speaking and listening, discussing and articulating, debating and defending, justifying and reasoning, are all part of the D&T experience.

However, these are only part of the process as language (as part of literacy) also entails concepts such as facial expressions, body posture, movement and tone of voice all of which can be explored through the products that we engage with.

The truth is that the subject is so heavily literacy-based that without good literacy skills being developed pupils will struggle. This requires the teacher to exemplify, signal and model the language skills that they expect from pupils as well as encouraging pupils to read appropriate design-related literature to extend both their technical and creative vocabulary.

A simple example would be, when designing a product, make sure you get pupils to capture, through a list of adjectives, what they are trying to achieve. These could be related to key features such as shape, colour, texture and feelings. For example a sensuous, organic, robust, reassuring, intuitive design for an MP3 player would give a clear set of expectations that the product could be evaluated against.

If you want to improve pupils' D&T capabilities then improve their literacy skills!

It has long been argued that the curriculum is too polarized and that too often children do not transfer or apply their learning within different areas of the curriculum. This is particularly unfortunate in D&T as it is a subject that should be feeding off all other subjects.

Most recently the biggest influence on D&T is the STEM (Science, Technology, Engineering and Maths) agenda. STEM is an attempt to coordinate the range of activities that operate under the different headings in order to provide a more focused approach to developing our economic competitiveness and industrial future. D&T does, however, have equally strong links with art and design education and again this provides worthwhile opportunities for developing collaborative activities.

With collaboration, however, there is a danger that D&T becomes merely the 'making' area and it is important when developing links with other subjects that you are building upon a range of good practices and disciplines which other subjects do not have. Therefore, any activities should be building upon the creative and process methodologies that are within D&T, such as the ability to take action, while drawing upon other subjects' approaches and methodologies.

Finally, such links are not merely to teach better D&T, better science or better art, but more importantly are about developing complementary dispositions geared towards the global dimensions of enterprise, creativity, cultural understanding and diversity.

Activity: Look at other subjects' National Curriculum guidelines to see where the most natural links with your own preferences fall, and look to see how these links can be formalized through joint cross-curricular activities.

LINKS WITH OTHER SUBJECTS

YOUR LIBRARY

Your library is your portrait.

I heard the story recently of a teacher being shown around the school and being shown the library. The teacher enquired that she thought it was a beautiful library but how would you be able to get more than 70 children inside? At this point the person showing around replied that it was the staff library! There are a couple of important points here.

Firstly, the old adage that 'you are what you read' does have some resonance, and a quick list of the last five books you have read or are reading might give some indication of what kind of person you are and the direction you are going in terms of your thinking.

Just to extend this point a little further – at a conference a few years ago the D&T audience were asked by the keynote speaker what they read related to D&T. There were some replies of 'I'm too busy to read', but generally there was a frightening silence. The speaker, a well known designer, was really quite upset at this, and perhaps rightly so: if D&T teachers are not reading D&T-related materials, then how can they possibly keep their pupils up to date with the latest ideas and thinking?

I think if the speaker had asked what educational materials teachers were reading the response would have been the same as above – deathly silence. Of course we can adopt a head-in-the-sand approach but this doesn't help you or your pupils. So the first point is that there really has to be an up-to-date collection of books that you engage with, whether these are at home, in the staffroom or the local library. I make the distinction of books as opposed to emails, webpages, etc., as they are different disciplines which you engage with in different ways.

The second area is: what are your pupils reading? If they are merely engaging with textbooks then this is unlikely to be of real benefit. They may help in developing answers to questions but they are unlikely to provide the breadth that is needed for genuine engagement in D&T. Subscribing to a range of good

D&T magazines (not just school-based ones) is really important.

Finally a few questions that should prompt your thinking further:

○ How good is your department's D&T library? Do you have one?
○ How big/good is the D&T section in your school library?
○ How big/good is the teacher section in your school library?

CITIZENSHIP

Citizenship is a compulsory subject for all pupils at KS3 and KS4 and many schools have adopted a cross-curriculum approach for delivering it. Regardless of this, D&T has a rewarding part to play in discussing the role of the citizen from an informed perspective relative to the designed and manufactured world.

A key part of this is that pupils become informed so they can participate, be critical and be active in decision-making within their communities. This means being aware of the made world and how decisions relating to new products, systems and environments have considerable impact upon the earth's resources, the workers involved in the production of such objects and the communities within which such products exist. Such discussions do not have to take place in a detached and sterile way. Instead they can provide rich stimulating contexts for discussion and decision making, design and action taking.

An example might be a discussion of 'design against crime' which links to a design council initiative (www.design-council.org.uk) that considers the social as well as the design issues associated with crime. Considering such contexts for design provides strong links to values and social, moral, spiritual and cultural issues as well as the emotions related to design, crime, consumers and producers.

This particular example would also link to the *Every Child Matters* agenda which relates to staying safe. It is no coincidence that such stimulating contexts are very successful at dovetailing many different agendas in a natural rather than contrived way.

A key government aim is that every child, whatever their background or their circumstances, has the support they need to be successful, and participative and active citizens. As part of this are the five key headings (BEAMS) around which the ECM agenda revolves.

D&T, like all subjects, has a key role to play and below are some ideas for linking to each of the areas:

○ **B**e healthy – this clearly has links to the healthy eating and food agendas but also links to emotional wellbeing.
○ **E**njoy and achieve – this links with engagement through a high quality D&T experience.
○ **A**chieve economic wellbeing – this can be delivered through recognition of sustainable communities through design, development and enterprise.
○ **M**ake a positive contribution – this links with the SMSC and Citizenship agendas.
○ **S**tay safe – this clearly links with activities such as the Design Council's 'design against crime'.

It is important to avoid the overly tenuous and contrived approach by looking for genuine impact and sustainability, and recognize that *Every Child Matters* represent a multi-agency approach to pupils and not just a curriculum response.

EVERY CHILD MATTERS

There are two key strategies, which should have a significant influence upon KS3, that offer clear frameworks for enhancing the quality of teaching and learning. Both are based around EPET: Expectation, Progression, Engagement and Transformation.

Strategy One relates to the Foundation Strand subjects, of which D&T is one, and contains the following packages: Planning and assessment – assessment for learning in everyday lessons; The formative use of summative assessment; Planning lessons; Teaching repertoire; Questioning; Explaining; Modelling; Structuring learning; Starters; Plenaries; Challenge and Engagement; Knowing and learning; Principles for teaching thinking; Thinking together; Reflection; Big concepts and skills.

Strategy Two relates to the designing strategy for D&T, based upon: The Framework – the vision; Planning as a team to teach the Framework objectives; Planning to teach the Framework: medium- and short-term planning; Teaching the sub-skills of designing; Transition – building on prior work about designing; Creativity; Starter activities; Plenaries; Questioning, Thinking together when designing; Modelling designing.

The first step is to audit the extent to which these activities are reflected in your learning environment, as when combined together these strategies provide valuable and powerful learning opportunities for all pupils, and as such, all teachers should be familiar with the contents which are readily available.

Structuring
the learning

A key part of supporting pupils' learning is the three 'S's: Signposting, Signalling and Scaffolding. The terms are partially self-explanatory and relate to Vygotsky's concept of ZPD (Zone of Proximal Development).

'Signposting' is indicating the way the learning is going to pupils, and stating why the learning is taking place. Examples are statements such as 'What we are learning is . . ." and 'The reason we are learning this is . . .' Such signposting is important for those learners who need to see the big picture before they can learn.

'Signalling' is the signals sent to pupils to alert them to key parts of what they need to know, and this can be both through verbal and body language. Although a teacher is potentially giving out a lot of information, it is important that those critical parts of learning – perhaps those parts of learning that other learning relies upon – are signalled to pupils. This could be by physically indicating while speaking, such as 'It is crucial that you understand this so make sure you are listening as I am going to ask you some questions afterwards'.

Finally, 'Scaffolding' is beginning on a firm foundation of existing learning and then building higher by extending the learning, enabling learners to reach beyond their existing competencies to explore new understandings and skills. Such learning is central to ZPD and is the basis of progression.

Activity: Next time you are involved in appraisal, peer review or any observation of your lesson is taking place, ask the observer to monitor the extent to which you are using the three teaching tools of signposting, signalling and scaffolding.

CHUNKING UP THE LESSON

It is increasingly being recognized that small chunks of learning represent more powerful learning and, therefore, 'chunking up' a lesson into a series of parts is likely to both increase pupil performance and improve pupil behaviour.

A key part of this is recognizing how big the chunks should be.

On average a learner aged around 11 can be expected to concentrate for 8 to 10 minutes, with this increasing, hopefully, as children get older. However, at different times of the day and at different times of the year this will vary.

This means that an average 60-minute lesson may need to have six 10-minute episodes and as such, planning should be based upon six chunks of learning, each building upon the previous one in a series of increasing steps.

The most effective way of understanding this is that after about 10 minutes, pupils are likely to be just about to go off task – this is when the fidgeting and restlessness starts. At this point, it is crucial the teacher intervenes and signals and signposts the next episode. Do this too early, however, and it upsets the rhythm and flow of learning – timing is everything!

In practical lessons pupils are often able to stay on task for longer, but the 10-minute rule is also useful to remember in these lessons as a means for dropping in some further learning. This could be in the form of a mini plenary or reorientation activities where pupils share learning and when the monitoring of progress can take place.

THINKING ABOUT SACK

The extent to which you plan a lesson is purely a professional decision. Your planning may simply exist in your head, or be written down for evidence and ease of use. Some schools expect all teachers to submit written lessons for the week ahead on a Monday morning, while other schools only ever see lesson plans from trainee teachers and when OFSTED visit.

Regardless of the method it is worth considering the balance of the lesson from a Skills, Attitudes, Concepts and Knowledge (SACK) approach.

There are two main benefits to this. Firstly, by identifying whether a learning objective is based upon a Skill, Attitude, Concept or Knowledge will determine how you assess that learning objective. For instance, if your first learning objective was based around a skill, then it is unlikely that this could be assessed through a piece of written work or a test, but more likely it would be assessed through observation of the application of the particular skill. Likewise, it is unlikely that you could observe knowledge being learned, however, you could assess the application and recall of knowledge through a test, written homework, etc.

The second reason for identifying whether you are delivering a Skill, Attitude, Concept or Knowledge relates to balance. A lesson with all the objectives relating to a skill may indicate an imbalance in learning. Although it might be tempting to consider a good D&T lesson being heavily skills-based, such a lesson – unless a one off – may provide pupils with a poor learning diet. A mixture over a period of time of Skills, Attitudes, Concepts and Knowledge is likely to be more appropriate and beneficial for learners.

All I really needed to know I learned at nursery!

It's not so much what you know but how you proceed when you don't know. For many years, education has been based upon a banking mentality. The teacher deposits something in the learner hoping that it may gain some interest. Such an approach has significant limitations: the appropriateness of what is being learned and (continuing the metaphor further) the transferability of the currency – particularly using the deposited information in different contexts and at different times.

An alternative way of conceiving knowledge acquisition is using a manufacturing analogy of 'just in time' and not 'just in case'. The basis of this principle is that there is little point of having masses of Skills, Attitudes, Concepts and Knowledge 'just in case' you might one day use it.

Therefore a more appropriate approach would be that you access knowledge in real time or 'just in time' for when you need it. Ultimately what this represents is knowing how to learn – learning to learn, and most importantly how to proceed when you don't know something. This could also be conceived as 'catching' learning, and environments have to be created where children feel comfortable to catch the learning.

Such environments already exist in many nursery schools where children are constantly in a state of experiential learning and are learning together without even knowing – they are literally catching the learning when they need it!

CATCHING LEARNING (1)

CATCHING LEARNING (2)

Increasingly, research is showing us that 'catching' learning is a useful metaphor for reconceiving a learning environment. In essence this practice lies at the heart of D&T. It is (or should be) a 'just in time' environment, where children are working on problems that they won't initially have the understanding to reconcile until they immerse themselves in the topic. Such learning is incredibly rich, powerful and highly motivating, and represents learning for progression and not just learning for the sake of learning.

In this environment the teacher's role is critical in facilitating and encouraging such dispositions. It is important that the teacher models similar behaviour, such as talking about how they have found learning difficult at times and what they have learnt recently. Obviously there are certain things that the teacher must know, but it is equally important that the teacher shares that they are learning also.

There is a reason for this – we have so-called 'mirror neurons' in the cortex of the brain that automatically prime us to mimic what we see others doing. In other words we are subconsciously 'catching' the learning, and if we have sufficient emotional engagement with the activity then the more effective this will be.

If pupils are learning and the teacher is showing enthusiasm about how they learn then the learning disposition is likely to be all the more effective.

There is a story about an Ofsted inspector who asks a head of department to describe the Year 7 keyring project that his pupils were working on:

'Well, they cut a piece of acrylic, drill a 4mm hole and polish the edges.'

The inspector then asks him to describe a Year 10 clock project that his older pupils are doing:

'Well, they cut a piece of acrylic, drill a 10mm hole and polish the edges.'

You will get the point from this example, which could be applied to any of the areas of D&T, that although something may be bigger, it does not represent any progress in terms of demand between Year 7 and Year 10 – something which is simply unacceptable.

Of course, the Year 10 group may appear to have developed over time, but unless progression is built into, and mapped against, activities in the form of the development of Skills, Attitudes, Concepts and Knowledge then pupils are unlikely to progress.

The tight prescription by the teacher in this example has curtailed opportunities for pupil progression in terms of designing as, although there may be a token opportunity for embellishment (often restricted to basic choices about shape and colour), the reality is that the teacher has determined the outcome, and in doing so has hindered any real pupil progress.

Therefore, any teacher planning must not only consider the development of SACK from year to year, but also project to project across each individual year.

PROGRESSION – ARE YOU SURE?

ALL THIS SPINNING IS MAKING ME DIZZY (1)

Rotation, circus and carousel – you get the picture. We are talking about the means of delivering the different material areas at KS3, which usually involves pupils experiencing five different areas – each approximately eight weeks long with five different teachers across the year.

Unfortunately, such arrangements have received considerable criticism: the major one relates to the lack of continuity and progression that pupils are experiencing in constantly switching from one teacher to another, meaning that no one ever gets to know the learners effectively. The main criticism, however, relates to the lack of progression across the year: for example, the learner who takes food as their last module in the year will not experience anything different from the learner who was taught the module at the start of the year – there has been no progression or added value across the time period.

The problem is often that the modules are designed as discrete entities – although this means that they can be taught at any time across the year they also predetermine pupil abilities and therefore cannot build upon previous experiences.

Activity: If using a rotation, circus or carousel arrangement, it is important that you plan each activity across the year to ensure that each has added value and builds incrementally upon the previous experiences.

Generally there are several alternatives to the traditional rotation systems previously outlined.

Firstly, there is a model where the teacher is up-skilled to enable them to deliver all the areas of the rotation; although pupils will still go through the different modules, the teacher will move with them, and can then build in progression based upon previous knowledge of a pupil's performance.

A second alternative is that the pupils work in parallel streams and therefore might have one teacher for materials and ECT, with a different teacher for food and textiles, thereby breaking the subject into two chunks rather than five. Graphics in this situation would be considered generic to support each of the material options.

A third and more radical idea is that the predominance of the material areas is less apparent and programmes of study are delivered through genuine D&T experiences where pupils don't work through predetermined projects, but instead work through contexts. In such a situation a pupil may be set a series of design briefs at the start of the year which they can work on to a greater or lesser degree. Most importantly, in this scenario, the pupil is managing several projects in parallel through their own management skills rather than working through a series of premeditated tasks.

Whichever system you choose there have to be opportunities for ensuring progression both across the year and from year to year.

ALL THIS SPINNING IS MAKING ME DIZZY (2)

GROUP WORK – A COMMUNITY OF LEARNING

Moving from the sage on the stage to the guide on the side . . .

Working in groups is something that it is quite uncommon in many D&T lessons yet group work can provide real opportunities for fulfilling teaching and learning.

There is a temptation to think that the most effective learning occurs in isolation, with a teacher standing at the front of class commanding and instructing the listening pupils, particularly if you were taught in this way. However, although this is one method of teaching (which does not necessarily imply learning), it is not always the most effective means of developing a breadth of learning capabilities.

Moving away from teaching from the front of the classroom and letting pupils take responsibility for co-constructing their learning in groups provides valuable opportunities for pupils to develop their communication, interaction and negotiating skills, while also learning new concepts, attitudes and knowledge. Such practices require the teacher to take on a new role, and instead of directing and instructing they take on a guiding and facilitating role.

One of the reasons which is often cited for not promoting such communities of learning is that it makes the task of assessment difficult. Although there may be some validity in this argument when related to specific individual assessments, perhaps particularly diagnostic assessment, the truth is that group assessments, as in peer assessment, provide equally rich, group-orientated practices. Next time pupils are designing, try and stand back and promote group learning through an activity such as 4×4.

Often teachers will say that during a lesson the pupils are learning about group work, however, what this often means is that the pupils have been loosely put into small groups to work out what 'group work' means for themselves. Although this approach may have some merits, particularly if pupils are allowed to reflect upon the group dynamic, the reality is unless pupils are taught to operate effectively as a group then it is unlikely that they will grasp the concept of group work.

When a group is set up for group working in your classroom you are in effect asking them to deal with two main issues. The first issue is the task that has been set. The second is that you are expecting them to engage with group processes and the problems of negotiation, discussion, management and communication. You are also signalling to them that the sum of the individual parts (pupils) should add up to a greater whole (group), otherwise there is little point in having a group.

To do this, here are five basic tips that may help:

1 Clearly signal the benefits and expectations of working as a group – the sum of the parts should be greater than the whole.
2 Signpost the key skills required, such as clarity of communication, negotiation, delegation and management.
3 Make sure the task is a worthwhile one and lends itself to a group activity.
4 Make sure you are clear on your rationale for selecting groups, e.g. friendship groups, ability groups, interest groups, self-selected or teacher selected. Regardless of the grouping it is essential that all pupils are encouraged to be active within the group.
5 Make sure that there are opportunities for reflection points throughout the activity to identify how effective the group work was.

GROUP WORK – FIVE TIPS

SHARING LEARNING OBJECTIVES AND OUTCOMES

In many classrooms, it is expected that teachers should share the learning objectives/outcomes with pupils; it is considered to be good practice. In fact, many schools will have a separate whiteboard at the side of the main whiteboard for writing this information.

Many teachers will put the information down but might not always be convinced of the benefits. In terms of research there is mixed evidence about the benefits of sharing the objectives and outcomes. However, it stands to reason that if you are going to share the outcomes then learners need to receive good information. There are two key points to remember for this:

Firstly, the objectives/outcomes must relate to learning and not just the 'doing'. So for instance, although an objective of the lesson might be to finish something/cook something, this does not necessarily imply learning will take place. So the learning must be extracted from the doing.

Secondly, a good way of writing an objective/outcome is to select an active verb to describe what the learner will be able to do as a result of the learning. When doing this, a really good idea is to link to Bloom's 'Hierarchy of Learning'. Using Bloom's taxonomy we are able to distinguish between low-level learning, such as knowledge or comprehension, and high-level learning, such as synthesis and evaluation. For example, if pupils are evaluating, then as a result of learning about evaluating they will be able to (choose one): Appraise, criticise, assess, argue, justify, defend, interpret, support, estimate, evaluate, critique, review, and write.

'Planning lessons is just for trainee teachers.'

Many years ago a statement like the above would be true, but in the last 10 years the increased accountability in teaching, and introduction of professional standards, has meant that the requirement to have documentary evidence for all lessons is more apparent.

There are, however, likely to be significant differences in the planning (or designing of a lesson) that trainee teachers do and what experienced teachers do, as trainees have to provide documentary evidence of the ability to meet the standards required to become a teacher.

Generally the minimum that any lesson plan should contain, whether for documentary evidence or simply to get through the lesson, are four areas. These are:

o Objectives/outcomes
o Assessment of objectives/outcomes
o Teaching column
o Learning column.

The use of objectives is discussed elsewhere in this book, however, the above points provide the key points/chunks/episodes of the learning. Essentially, for every objective there should be a teaching episode and a resulting learning episode accompanied with a corresponding assessment opportunity to check that what has been taught has been understood.

A common mistake is that the planner or lesson designer will start by considering what they are going to teach first. A much more effective way is to first consider what are the pupils going to learn, then how this can be assessed and finally consider what is the best way of teaching this to achieve the desired learning and assessment.

A further key point is to consider this as lesson *designing* rather than just planning. This means, as in all designs, that although you have considered a particular way of delivering the learning, you may want to reconsider it several times to get the best designs for the desired learning.

PLANNING AND DESIGNING LESSONS

The extent to which this is written down and presented will be very much dependent upon the school requirements, however, it needn't be onerous and can easily be achieved on one side of A4 paper.

In many schools it is considered good practice to share learning objectives for the lesson with the learners. It is really important that teachers get pupils to engage with these objectives as otherwise this becomes a meaningless activity which merely fulfils school requirements and wastes a valuable learning opportunity.

At the very least, pupils could be asked to read out what the objectives of the lesson are, but a more important question might be, 'Why are we learning about these objectives?' This means that pupils have to think about the context of the objectives within their own learning. Another common feature is to review the success in meeting the objectives at the end of the lesson.

A frequent problem with many objectives in D&T is that they are not based upon learning and are merely based upon doing. So for instance, 'Today you will complete X and Y' does not guarantee or imply any learning. However, if the objective is expressed as 'Today you will be learning X' then it forces the teacher to consider the learning rather than the doing. It is also important to remember that children are meant to be involved in learning through the doing, and if they are only doing (and not having clear learning) then it is difficult to be able to justify such an expensive and time consuming experience.

A simple way of carrying this out is considering what the pupils are learning, and what the difference in terms of their learning is between the start and the end of the lesson. If we can guarantee that pupils have learned either new Skills, Attitudes, Concepts or Knowledge (ideally a mixture) then this will make pupils more successful.

It has been repeated many times throughout this book that D&T offers a unique way of learning, namely learning through doing, by taking action. This process is very much at the heart of the subject – children engage in possibility thinking, and in doing so design and manufacture products, while learning about the associated materials, processes, technological and environmental issues associated with their ideas. Through this they consider the social, cultural, emotional, political and philosophical implications of their decision making.

A real danger is that such learning is often bypassed in order to simply achieve a product outcome – the appearance at the end of the product manufacture may appear the same whether or not learning has taken place.

Bypassing of learning is often due to the assumption that merely going through an activity and producing a product will mean that children automatically learn. Although it is true that children might learn, unless this learning is planned, crystallized, signposted and scaffolded for the learner it is less likely that the desired learning will take place.

Poor practice adopts a process of learning by osmosis, where it is assumed that pupils will purely and naturally be absorbing their learning from their experience when in fact they may not be. What is more, they may in fact be learning that opposite of the intended learning and generating their own misconceptions.

Central to avoiding learning by osmosis is teaching which clarifies the learning through the doing, which also indicates the outcomes from such learning, and finally which clarifies the assessments required for gauging the success of that learning.

Including all

GIFTED AND TALENTED

'Gifted and talented' remains an emerging and as yet un-robust term in many schools, as the tendency when selecting who is gifted and talented is to work on percentages rather than actual capabilities. Therefore, the average school works on the basis that five per cent of the student body – approximately 50 pupils – will be gifted and talented. Added to this, is that it is estimated that it takes 10,000 hours of practice to become really good at something, and as indicated in this book, a learner's secondary school experience in D&T may be a little as 1/100th of this. Consequently it is unlikely that you are going to see a fully-developed gifted and talented child in D&T unless they are gaining significant experience of the subject elsewhere.

However, there are general signs that we can look for with gifted and talented children, such as:

o bored and frustrated easily with the level of work
o often do not work up to their potential in the classroom and are sometimes seen as disruptive, hyperactive, inattentive or impulsive
o curious by nature and eager to learn
o perfectionist in their work which leads to stress and tension
o often work alone and have difficulties in group work
o feel and are socially isolated from their peers
o accelerated cognitive development – extensive vocabularies, high reading ability or an extensive understanding of mathematical concepts
o good listeners and express themselves well
o divergent thinkers and novel problem solvers
o prefer creative activities
o sophisticated sense of humour.

An important feature of helping gifted and talented pupils is that they do not know what progression in the subject looks like, and it is therefore essential that they are given opportunities to be extended through out-of-school learning contexts and competitions where they can mix with children of similar abilities.

Give a boy enough rope . . . and he'll come back with a dog!

72

The teaching of boys in D&T has received a lot of attention over the last decade, particularly as boys do not seem to be doing as well at girls when it comes to passing GCSEs in the subject. This has resulted in the appearance of a gender gap – a lot of people make a fuss about it, but there is no clear rationale as to why there shouldn't be a gap – either way!

However an important point to note is that it is not *all* boys that are not being as successful as girls. Although some boys may not be passing the same quantity of GCSEs it doesn't mean that they are not doing well in the subject. Another way of thinking about this is that boys are doing reasonably well despite an assessment system that seems weighted against them.

What we do know from research is that some boys prefer to work in different ways in the subject, and overly constrained activities throughout a project may impede their success in the subject.

The problem is, however, much more complex than this, and in my own research I have calculated four billion (yes, billion) possible combinations of factors influencing boys attainment in the subject. One area I am convinced about is that good teaching overcomes bad testing, and many of the ideas in this book will improve boys' performance and raise attainment.

Activity: If boys are not doing well in your school, then try to unravel the complex reasons for this. It is unlikely to be any one factor and wherever possible, in order to increase the chances of success, try to go for a variety of solutions rather than one big solution.

Two boys were arguing when the teacher entered the room. The teacher said, 'Why are you arguing?'

The first boy answered, 'We found a £10 note and decided to give it to whoever could tell the biggest lie.'

'You should be ashamed of yourselves,' said the teacher, 'When I was your age I didn't even know what a lie was.'

The boys gave the £10 note to the teacher.

Here are some (of the many) tips that may be worth considering when teaching boys:

○ Ensure that department planning is aware of gender issues, have a named person responsible for this, and plan activities which do not have significant gender bias in them, or at least be aware of any potential bias.

○ Use peer appraisal to observe and develop teaching styles to ensure there is not a gender bias in the teaching of boys.

○ Ensure consistency of sanctions to both boys and girls.

○ Make deadlines explicit to all pupils and ensure expectations are clear.

○ Develop compensatory (catch-up) activities.

○ Use alternate questions in lessons, e.g. boy then girl, then boy then girl, to ensure there is not a bias towards a particular group.

○ Know your pupils in your school, but do not limit your expectations by past experiences with similar pupils.

○ Tackle anti-learning cultures – show that learning provides choices.

○ Ensure there are examples of positive role models (e.g. photographs of successful past pupils).

○ Encourage parental support and understanding.

○ Make sure all pupils are clear about how they are assessed (see AfL, page 83).

○ Use affirmation statements, positive language and have high expectations of all pupils.

○ Use a variety of teaching strategies that adjusts with the activity, group, time of day and time of year.

- Engage pupils with activities that challenge and interest them.
- Education has to be considered as part of a continuum, the more you know about pupils' prior learning, the faster their progress will be.
- Recognize and reward the diversity of pupil achievement. All pupils have something to offer; often they will have capabilities in areas that are not always measured by our often crude and simplistic assessments systems.
- Think about the context in which you set pupils work. Some contexts will engage pupils whilst others will not.
- Remember, there are unlikely to be quick solutions, or one solution that fits all!

There is a story about a teacher who goes to see the headteacher to offer her resignation as the children in her class are completely out of control. Unfortunately the head has to deal with another problem at that time so asks her to wait in his office.

While in his office waiting she sees on the headteacher's desk her class list with some numbers next to her pupils' names – particularly high numbers are next to some of her most naughty pupils. She starts to think perhaps little John with a score of 147 isn't that naughty after all, and little William has a score of 143, while Rachel (who has been giving her lots of problems) must just be over-exuberant as she has a score of 150.

When the headteacher returns, the teacher tells him it didn't matter and went back to her class. A few days later the headteacher pops into the class and notices a much more relaxed atmosphere with the children seeming to be engaged and enjoying the lesson.

'Glad to see you have sorted things out,' he says, 'and by the way, here is your class list with your children's new locker key numbers on'.

If only this story were true! The point, however, is a pertinent one, and it is that our expectations often resemble our selective perceptions, and our pupils will often resemble our perceptions – we in effect choose what we want to see in pupils. Try and see the best in all pupils, as if you think of your pupils in negative ways they simply won't disappoint!

The term 'underachievement' has become something of a catch-all statement often used to capture the frustration that some teachers feel when dealing with certain pupils. However, the term is misleading, it has no validity and is often a metaphor for X is a 'difficult pupil to teach'. The problem is that labelling any pupil as an underachiever can lead to self-fulfilling prophecies; pupils will become what you think they are, so think positive.

Boys particularly have become casualties of the labelling process and are regularly labelled through the media as society's underachievers. Some boys may be difficult to manage, have bags of energy, be noisy and have untidy handwriting – whatever extra demands they may place upon you, they should not be placed at a disadvantage because they belong to a particular sex.

It is true that some boys are not achieving the same results as some groups of girls; however, this varies by age, subject, region, social class and assessment methods adopted. In fact, if you try to calculate the possible number of reasons affecting attainment it is literally billions, yet many teachers will often simply indicate a lack of effort as being the problem. What is often forgotten is that teaching boys is a more sensitive barometer of effective teaching; rather than labelling them as underachievers, it could mean that you need to examine your teaching style, assessment strategies and subject content. Labelling a pupil as an underachiever may in fact be labelling yourself as a poor teacher.

Boys are a vulnerable group who are more sensitive than they sometimes appear, and by examining our own methods we can raise their performance – more importantly, using the term underachiever in an educational context can only add to the problems.

DO YOU USE THE TERM 'UNDERACHIEVER?'

DIFFERENTIATION THE EASY WAY

For years, teachers in D&T have advocated that children naturally differentiate themselves by the outcomes that they produce. To a certain extent this is true, if learners are genuinely involved in open design and manufacture activities.

However, many activities are not completely open-ended, and therefore we need to think about the different levels of differentiation. What will all pupils be able to achieve by the end of the lesson? What, at a higher level, will most of the pupils be able to achieve? Finally, what will some of the pupils, possibly the most talented, achieve by the end of the lesson? It is also important to share such differentiation in terms of targets and expectations with pupils at the start of the lesson.

A range of strategies are listed below which provide opportunities to engage with pupils through differentiated activities, and it is useful to audit individual lessons to explore the different types and levels of differentiation employed.

Differentiation by task	Yes/no
Opportunities for group work	Yes/no
Differentiation by resource	Yes/no
Opportunities for pupils to teach others	Yes/no
Differentiation by outcome	Yes/no
Opportunities for pupils to have discussion	Yes/no
Differentiation by starting point	Yes/no
Opportunities for pupils to have ownership	Yes/no
Differentiation by self-start	Yes/no
Opportunities for self-start	Yes/no
Differentiation by support	Yes/no
Opportunities for pupil to define outcomes	Yes/no
Differentiation by homework	Yes/no
Opportunities for different outcomes	Yes/no
Differentiation by question	Yes/no
Opportunities to use different creative methodologies	Yes/no
Differentiation through visual, auditory or kinesthetic learning	Yes/no
Opportunities for support and extension	Yes/no

*Does an Eskimo give his children fish? Or does he
teach them how to fish?*

In the example above, which is in the child's best
interest? What is the equivalent in schooling?

Education is allegedly changing to a personalized
learning approach. What this means in practice is largely
unknown, however, there are some key points that are
considered to be part of this:

○ Developing and promoting autonomy
○ Modelling of involvement and taking responsibility
○ Up-skilling pupils to take greater responsibility for
  their own learning.

This includes teachers understanding themselves and
pupils understanding themselves (their learning/thinking
styles), managing risks, managing failure, etc.

This also includes:

○ increasing peer-to-peer work
○ increasing group work opportunities.

Unfortunately, many of these activities are alien to both
teachers and pupils. In D&T this might include:

○ pupils designing in different ways
○ pupils making in different ways
○ pupils working in different contexts
○ pupils working alone and in groups
○ pupils completing parts of projects – instigators,
  finishers, etc.

By having a rigid formulaic approach in D&T, we are
alienating a significant number of children. We are also
telling them they are 'not very good' in D&T, when in
fact they may be!

PERSONALIZED LEARNING IN D&T

GHOST CHILDREN

Like the child in the film *The Sixth Sense*, I sometimes feel like I see ghost children. These are the children who wander the corridor, who no one knows, who don't cause problems, but don't set the world on fire. They are neither memorable nor influential, but they are there in every class. Not sitting at the front or at the back, not answering too many questions or too few – going with the flow. You sometimes don't know they are there.

In any class, 80% of your time is often taken up by a 20% minority of either very bright or very naughty (or combination of both) pupils. The ghost children are the ones in the middle who just 'get on' and who seem to get a raw deal. However, the ghost children are your potential high flyers who need to be challenged.

The key to this is spotting your ghost children, and this is easy to do – go through the names in your register. The names that you cannot put a face to are your ghost children!

Essentially, the point is that some children will simply try to go through the school day without causing too much fuss – this is not always in their best interests, although it may be convenient for both parties. To overcome this, monitoring systems have to be in place so that all pupils are engaged, whether it is through the use of rotas for handing out work, reading a passage from a book, or through a question and answer session – it is important that all pupils are engaged and that a silent minority don't disappear!

There is an assumption that schools are where pupils learn, not teachers, and while you are reading this you might want to consider the quality of the provision for maintaining your education in your school.

We are beginning to understand that learning is almost like a catchable virus (mirror neurons), and that habits of mind are almost contagious. Therefore, if teachers are learning, it is likely that their pupils are also learning.

Put it another way – do you expect your doctor, mechanic, or financial advisor to be up to date in their field – almost unquestionably yes. However, in education (apart from a few inset days) learning is often considered something of a 'busman's holiday' and something you only need do occasionally. Yet research is informing us all the time about new developments in teaching and learning, and so much of it is more accessible than ever before. Although the demands of time are great, it is critical, for the benefit of the pupils, that all teachers remain active learners.

One of the most effective ways is for teachers to undertake research, often called action research. Such research is incredibly beneficial, as it implies action is being taken not only to become more informed but also to improve practice. It's the ultimate win–win situation!

Activity: There are many funding opportunities for teachers to carry out research in schools, however, the first step is to build up a bank of literature related to your area of interest, so that when you apply for funding you have already carried out your initial analysis and investigation.

TEACHERS AS LEARNERS

Assessment

Teacher: 'Did your father help you with your coursework?'
Student: 'No, Sir, he did it all by himself.'

THINKING ABOUT 'LEARNING HOMEWORK'

Typical scenario: it's the end of the lesson, everything is in a rush and the final instruction from the teacher is '. . . and your homework is to finish off anything that you didn't complete in the lesson'. Fair enough, you might think?

Well there are three fundamental flaws.

Firstly, homework has always been given at the end of the lesson, but this doesn't make it right. If you think about it, at the end of the lesson is probably the worst time to deliver the homework as this is when pupils are most likely to be distracted, with their minds already on the next lesson. A good place to deliver homework is at the start of the lesson, as throughout the lesson the teacher can signal the relevance of the learning to the completion of the homework task.

The second flaw very much relates to why we give homework. If homework is viewed as punishment for a lack of productivity in the lesson, as in the example above, then a real learning opportunity is being lost. Think about it – the pupils who are comfortable with the work are not being stretched, while those who could not complete the work in the lesson are unlikely to be able to complete the same work outside of the learning environment, and will consequently fall further behind.

The final flaw is in the name itself – 'homework' – which is synonymous with the second flaw, namely that a lot of homework is not about 'learning' but about a bit more of the 'doing' that you have already studied. It doesn't really extend learning. However, re-labelling it as 'extended learning' at least forces the teacher and pupils to consider its potential as a vehicle for extending learning in a different location.

The solutions to the flaws identified are resolved by providing clear differentiated 'extended learning' at the start of the lesson, with obvious signals throughout the lesson to show pupils that what they are learning (not just 'doing') relates to their 'extended learning' work.

There is a story about a deputy headteacher in Newcastle-upon-Tyne who told one of his students to 'pull the other one' when he said he wanted to be a footballer. The same deputy told another guitar-playing student that he would 'never get anywhere playing that kind of stuff.' Apparently both Alan Shearer and Mark Knopfler both did rather well!

The truth is that the pupil assessment is not an exact science and often we can be very wrong, but this largely depends upon the disposition we adopt.

A starting point is to think of your position and how you see your role. So do you see yourself as:

A A Distributor – you distribute grades to children. So your role is simply to give out grades based upon your experience.
B An Allocater – you allocate children to different social groups (not based on attainment). So, one won't achieve much, one's a high flyer, one's a loser and so on.
C A Hiker – no matter who, you aim to hike the child up. You work out where a child is in their learning and take them from one step to the next, constantly providing them with the next challenge.

Hopefully your answer is C, but there are lots of As and Bs in the teaching profession, what we call 'schoolers' rather than 'educators'. The difference is that a schooler doesn't add value to the child, and sees their role as clarifying a child's particular station in life. The educators are the ones who believe that children have unlimited potential and who are always providing the next challenge. Central to this process is assessment, and whether you see assessment as a tool to tell children where they are and hope they get better, or as a tool to tell them where they have just left and where they are heading.

THINKING ABOUT SUMMATIVE ASSESSMENTS

Summative assessments have had something of a bad press over the last few years, with so much emphasis being placed upon formative assessments as the way forward. However, summative assessment offers something different for the teacher, and one idea is to use it to develop effective monitoring and compensatory activities.

The simplest way to do this is for teachers in D&T to decide what generic capabilities they value most as a department. This could be accuracy, design ability, time management, communication skills, etc. At the end of each activity, if these generic areas are summatively assessed by the teacher (and self-assessed by the pupil) and secured in a spreadsheet or database then over time a summative pattern will emerge for each area and each pupil.

This information is incredibly valuable for a department to analyse the extent to which courses need to be adapted to improve the generic capabilities of pupils. However, more importantly, it is beneficial at pupil level – if a particular pupil is consistently scoring low on 'accuracy,' then rather than putting that child through more activities where their weaknesses are merely going to continue, you would instead put them through a compensatory activity with other children focusing upon improving accuracy. Such principles are at the heart of the personalized learning agenda, and offer real diagnostic ways of improving pupil performance.

Imagine going to the doctor and her telling you that she couldn't say what was wrong with you, or how to get better, but she could tell you how bad it was on a scale of one to ten. I think you would be largely unsatisfied. It doesn't take long to draw a similar comparison to summative assessments in school, where we can tell a pupil they are seven out of ten but we are not saying how to improve or exactly what the seven out of ten represents.

The key to Assessment for Learning (AfL) is that it is driven by learning, and, therefore, there has to be transparency in the process. So, when feeding back progress, we can report back that the work would be Even Better If (EBI) more attention was paid to X or Y. An essential feature of AfL is that pupils, with guidance, learn to take responsibility for their own learning and assessment, as any system which builds a high dependency upon others for feedback (and then removes that feedback system when the child leaves school) is merely short changing the pupil.

THINKING ABOUT ASSESSMENT FOR LEARNING (AFL) (1)

THINKING ABOUT ASSESSMENT FOR LEARNING (AFL) (2)

There are of course many components to AfL, some of which are referred to in this book (such as the 'no hands rule', page 100). My top five tips for AfL in D&T are:

1   Target pupils. Keep a list of individuals who you are going to target in each lesson to ensure that you have one-to-one dialogue with them about their progress. By targeting and recording you are ensuring that you are purposefully engaging with every pupil over a period of weeks, ensuring that all receive high-quality formative feedback, rather than randomly selecting pupils.

2   Peer assessment. There are few areas of teaching more powerful than pupils engaged in dialogue with each other about their work. By giving pupils the tools to peer assess (not mark) each other's work, they are not only providing each other with essential information, they are also employing high-level cognitive skills by articulating their thinking, which is essential to their own development.

3   Model and share. Teachers should model good work in order to share with others what they mean – this does not mean telling pupils exactly what you expect, but instead drawing out the best features of exemplar work for all to see and learn from (including parents and other colleagues).

4   Share the assessment criteria. When giving homework, share the criteria that you are assessing against. When starting a new project, share the criteria that you are assessing against. When students are working in groups, share the criteria that you are assessing against. Get the picture!

5   Plan good questions. The art of good assessment is to find out what pupils don't know rather than what they appear to know. Therefore, the planning of good 'thinking' questions will help you identify the limits of learners' understanding.

Teachers on average ask around 350 questions a day. This represents roughly 16% of classroom time devoted to questions: questioning is a large part of classroom routine. However, when analysed, approximately 60% of all questions are low-level recall-the-facts type questions, while only around 20% require pupils to actually think. This leaves the remaining 20% as procedural questions such as 'What are you doing?', etc.

It is therefore important to think about questioning some time prior to entering the classroom. Questions can be about revealing and challenging misconceptions that pupils have – a good question is not one that a pupil necessarily gets right, but is instead one that reveals something about their thinking.

To improve the quality of questioning here are five simple tips:

1 Plan your questions before the lesson.
2 Preferably use 'why' as opposed to 'what' at the start of a question.
3 Get pupils to plan questions for each other and the teacher.
4 Planned questions should be linked to the objectives of the lesson.
5 If a pupil gives a wrong answer, respond with another question.

Remember, Bloom's Taxonomy provides an excellent structure for questioning, starting with low-level knowledge recall answers, such as 'What did we do last lesson?', to synthesis and evaluation type questions, such as 'Why do you think we learned what we did last lesson?'

THINKING ABOUT QUESTIONING

'LOOK – NO HANDS!'

Do you remember being at school and trying to be chosen to give an answer to the teacher's question? Putting your arm high into the air while trying to remain seated, stretching every sinew just so you could get your arm an inch higher than the person in front, just so you might be called to answer the question?

Did you ever look around at some of the other pupils who might not be trying so hard? Or perhaps you were the one who sat back, knowing the teacher would pick one of those pupils whose seams on the armpits of their jumpers were about to explode due to the forces being exerted from their desperate extension of their arms?

One of the most misunderstood strands of the Assessment for Learning strategy is the use of the 'no hands' rule when asking questions. It is a move away from random participation in the lesson, where pupils put their hands in the air offering to answer a question while others sit back comfortably knowing they won't be selected. The 'no hands' rule creates a climate where everyone is encouraged to participate, and where everyone is expected to be able to contribute to the answer in some way. Part of this involves the teacher planning some really good thinking questions, and another part is creating a waiting time to allow all pupils to think about how they might be able to contribute to the answer. Although not all the pupils will get the right answer, often answers that reveal a misunderstanding provide good feedback for the teacher to be able to correct any misconceptions that the pupils might have developed.

Another feature of the 'no hands' rule is the concept of targeted questions, where the teacher considers before the lesson who they might engage in a question and answer session.

Planning and targeting questions to individuals means that those quiet 'ghost' pupils who often escape being asked questions are systematically included and monitored in the question and answer sessions, as well as the teacher gaining valuable assessment information from their answers.

The wider classroom

*Aerodynamically speaking, bees shouldn't be able to fly –
it's just that they haven't been told.*

LEARNING STYLES – THE TRUTH

Learning styles have become something of a boom industry in schools, with many companies generating significant sums of money promoting a particular type of learning system aimed at improving pupil performance. The truth is that we all have learning preferences (often very complex preferences), and adopting a wider range of teaching techniques may improve learning. However, trying to reduce this into a package or particular style is equally likely to disenfranchise as many learners as it manages to engage.

The debate as to whether you teach to a pupil's particular strength or style, or whether you develop a pupil's breadth of styles and work on their weaknesses has never really been fully developed. Some schools have persisted in focusing on learning styles with real commitment, and in such environments the very fact that there is such a commitment to, and dialogue about, learning is sufficient to raise pupil performance. However, the concept of assessing, for example, a pupil's dominant intelligence (as in the theory of Multiple Intelligences), and then teaching to the preferred intelligence/style seems limited. Whether such intelligences exist in the first place is a source of debate, and, secondly, whether these intelligences are fixed is also open to question. By constantly feeding a dominant style we may in fact be disabling pupils' all-round capabilities.

The most commonly talked about learning style in schools is visual, auditory and kinesthetic preferences (VAK). Again, this takes a common sense approach (as the areas broadly relate to our various faculties) and suggest a variety of teaching and learning approaches. For example, after watching a video clip, you might then have some role play, then a discussion, which all makes for a more dynamic and engaging lesson than if you simply just stand at the front and talk. However, this is hardly revelatory.

What learning styles do offer is a way of simplifying the complexity of teaching and learning, and if they provide a useful stepping stone for teachers to raise their awareness of the diversity of learning, then they can be viewed positively. However, beware – they are not a panacea.

Behaviour management is complex and a key concern for many teachers. This is perfectly natural, as how many of us are naturally equipped to manage large numbers of pupils who don't always want to do what we want them to?

An important part of behaviour management is how you conceive the context. Firstly, if a pupil is determined to ruin a lesson then it is highly unlikely you can prevent this, but there is a lot you can do to prevent it escalating. Choice is a key part of this – by offering pupils choices you are letting the decision making rest with them. Therefore, instead of saying 'I am telling you to stop that now!', if you say 'You can either stop that or you can be removed – you decide', this immediately places the burden of responsibility on the learner.

Secondly, behaviour management implies dealing with the *causes* of the bad behaviour, not just dealing with the results. For example, most controllable bad behaviour results from boring activities, unsupervised activities, inappropriate room layout, uncorrected triggers, lack of signalling by the teacher, and so on. Overwhelmingly, pupils involved in bad behaviour prefer their teachers to be clear and consistent about their expectations (what they can and cannot do) – it is the grey areas that often cause the problems.

Finally, a key feature of behaviour management is remembering you are often dealing with a tiny minority of pupils for a tiny minority of the time, although this takes a disproportionate amount of thinking time and energy, resulting in a skewed perception of the experience for both teachers and pupils. Therefore, re-conceptualizing is important – it is not all about catching pupils misbehaving, but also about catching pupils being good. Praise for good behaviour should significantly outweigh the criticism for poor behaviour: when you get the environment that you want, make sure you praise in a positive way and wait for the warm glow!

BEHAVIOUR FOR LEARNING –
CATCHING THEM BEING GOOD

I am going to give you a question that you should have a go at trying to work out what the answer is. The question will be very difficult to answer unless you know the rules of learning behind the question. Do not read the solution until you have spent at least five minutes trying to work it out. More importantly, try to reflect how you feel not knowing the answer and not knowing how to proceed.

$$L+M = 6$$
$$T+Y = 5$$
$$K+H = ?$$

Did you spend some time getting frustrated? Feeling lost and not knowing how to proceed? If I tell you the answer is calculated by counting the straight lines in the letters, and that the answer is 6 – suddenly a sense of order comes in to our thinking, 'Aha!' A simple 'learning of the rules' means that we can suddenly answer any question of this type, when five minutes earlier our brains were in a state of paralysis trying to figure out the answer, and we may also have been questioning our own ability.

The point to the exercise was to reflect on what that feeling was like, so that if we magnify that feeling several times over we understand how children feel when confronted with new learning.

Therefore, whenever we are presented with new learning, it is really important to break it down, if possible, into small chunks of rules that suddenly create the 'Aha!' moment.

HOOK AND 'WIIFM'

Do you ever hear a tune on the radio on the way to work and the tune stays with you all day? Well, the bit that you will probably be playing over and over in your mind is the 'hook' of the song. This bit is designed to capture your interest and is used to sell the song.

The same theory can be used in teaching and learning. When planning your teaching it is useful to think of the hook that is going to capture pupils' interest – it should be catchy and easy to remember. This might be the key point of the lesson, a particular formula or a point of reference. It is unlikely that it will be in the form of a tune (although there is no reason why it shouldn't be!) but you should be able to reduce a set of principles into one catchy line which makes it easy to remember. For example: 25, 2, 16, (keep saying it) 25, 2, 16 (and again) 25, 2, 16. This of course refers to the 25 transistors, 2 diodes and 16 resistors that can be found in a 555 timer chip.

Another way of engaging pupils is through 'WIIFM'. This means: 'What's In It For Me?'. So although 555 timers might be really interesting for the teacher – What's In It For Me as the learner? If the only reason we are teaching something is so that pupils can pass exams, then this often isn't enough for many pupils. However, if we can engage them by providing information that relates to the real world, and where they may unknowingly be aware of the topic in its real context, then we are more likely to be successful in getting the message across. So for instance, 555 timers in the real world can be used to keep premature babies alive – the pulse from the timer can be used to regulate a ventilator to help a baby breathe. A component that only costs about 15 pence can be critical in keeping someone alive. Keeping the learning engaging and snappy really does work!

PUPIL VOICE

Pupil voice = 'marking conversations'

Very much part of the *Every Child Matters* and personalized learning agendas is the theme of 'pupil voice', which ties in to the Ofsted programme of taking into account pupils' perceptions when undertaking school inspections.

For many, this goes against the grain of the way most adults were brought up – which was 'to be seen but not heard' – and many teachers feel vulnerable at the thought of pupils giving the 'low down' on their teachers. For some teachers there may even be a feeling of resentment as asking pupils their opinion gives them the appearance of an elevated status, creating a 'them against us' scenario.

However, there is a significant misconception about pupil voice – namely, just because a pupil says something it doesn't mean they are right. What it does do is provide a rich insight into pupils' perceptions, which can provide incredibly valuable information about their beliefs and misconceptions.

In essence it lets us find out quickly what otherwise may go unknown. An essential part of this is the concept of 'partners in learning'.

In D&T, a simple way of capturing pupil voice is through the use of simple questionnaires aimed at capturing pupils' perceptions of their lessons, their projects, their homework, their resources, the type of teaching they enjoy and so on. Remember when considering the results from such questioning, don't automatically give it a status beyond what it is – it doesn't mean that pupils, whether saying something positive or negative, are correct. However, the results provide information that allows for a rewarding discussion at departmental meetings and during individual discussions.

An important part of pupil voice is that it is not merely a collection of questionnaires – there has to be some action and follow up, and although this might not be a radical restructuring of the curriculum, often it may be that there simply needs to be some feedback of the big picture to pupils about why things are the way they are.

Although it can sound like a cliché, when education is a partnership between the school, parents and pupils, it is a very worthwhile and effective way of ensuring pupil success. One of the biggest factors in this is challenging parents' misconceptions about D&T; while many will recognize what other subjects are about, often they will simply not understand what D&T involves. Worse still, they may think it is something that it isn't, and hinder rather than help pupil progress.

Therefore, while teaching pupils it is useful to also inform parents about what the subject involves. The simplest way of doing this is through information leaflets – whether these are your own or standard ones from the Design and Technology Association (DATA).

Another method is to provide hands-on opportunities for parents in the form of a parents' evening – perhaps showing the latest technologies that you might have. Also, every time parents are in the school it is a further opportunity for educating them – so each parents'/consultation evening/day is again an opportunity to have high-quality displays showing what the subject is about.

Almost certainly this is a two-way process, as there are sure to be some parents whose expertise can be tapped into as their career directly relates to the subject.

Finally, through newsletters and emails you can keep parents informed of local events such as exhibitions or days out related to the subject, where they can take their children as part of enriching their wider D&T experiences.

Education is changing dramatically: instead of a local context we are now thinking about learning in a global context, in a highly technological and complex society.

These are some of the ways that learning is being conceived differently:

From	To
Reactive	Creative
Stable	Agile
Instruction	Construction
Quality controlled	Quality assured
Content delivery	User-generated content
Fit into the system	Fitted for the student
Individualized	Personalized
National	Global
One to many	Peer to peer
Interactive	Participative
Curriculum-centric	Learner-centric
Teaching	Learning
Pieces	Projects
Piaget	Vygotsky
Mundane	Engaging

The way we feel about ourselves, and our ability and willingness to take on new challenges, is related to our self-concept and self-esteem. The bottom line is that if we think we can do something then there is more chance that we will persevere until we get it right – as opposed to the likelihood that we will give up if we think we are unlikely to be successful. In a problem-solving creative activity this is even more apparent, as often pupils have to deal with great uncertainties.

One way of addressing this is through BASIS models, which consider the development of self-concept through:

B = Belonging. Creating a sense of community with a group or team.
A = Aspirations. Encouraging pupils' aspirations – making sure they don't set their sights too low.
S = Safety. Making sure that pupils feel safe to make mistakes, to test their ideas or to ask questions.
I = Identity. Developing pupils' positive self image.
S = Success. Recognizing and celebrating success.

The best way these are mediated is through both action and language, so we have to show as teachers that we believe all pupils are capable and use positive self-affirming language to encourage all pupils.

Activity: Record your use of 'affirming language' in a lesson – this can be with a small digital recorder in your pocket. When you get time, play it back and mark in two columns the frequency of positive and negative comments, and begin to examine the amount of positive feedback you give. To be even more sophisticated, examine particular dialogue with individuals – although you may be positive to the whole group you might find your reaction to particular groups or individuals is less positive.

ENCOURAGING SELF-ESTEEM – BASIS

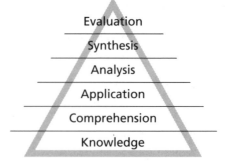

Most of you will be familiar with the above – Bloom's Hierarchy of Learning – however, how do you use it? The simplest way is to design learning activities around it. To do this we need to select the active verb from the corresponding levels and use these to steer the learning activity in the form of a 'signposted' learning outcome.

Therefore, if we want a low-level activity, we could ask pupils to label the parts of a diagram (knowledge). However, if we want a high-level activity then pupils will need to be able to critique a range of different ideas (evaluation).

Knowledge: define, label, recall, order, list, quote, match, state, recognize, identify, recite.

Comprehension: describe, discuss, summarize, paraphrase, report, review, understand, explain.

Application: assess, demonstrate, examine, distinguish, establish, show, report, implement, determine, produce, solve, draw, interpret, provide, use, utilize, write.

Analysis: analyse, illustrate, discriminate, differentiate, distinguish, examine, question, infer, support, prove, test, experiment, categorize, write.

Synthesis: Compile, categorize, generate, negotiate, reconstruct, reorganize, revise, validate, organize, plan, propose, set up, write, substitute, initiate, express, compare, modify, design, create, build.

Evaluation: Appraise, criticize, assess, argue, justify, defend, interpret, support, estimate, evaluate, critique, review, write.

One of the most important features of education is to teach learners that what they think they see or feel isn't necessarily the norm – it is in fact a highly filtered perception of reality. Therefore, any decision making that pupils make must be countered by the fact that others may perceive their ideas in a very different way. Although we think we know what someone else might think about our ideas and designs, unless we ask them we won't know.

Here is a simple example of how to illustrate this. Firstly, ask your class to count the number of Fs and the number of As in the sentence below, using only their eyes:

FINISHED FILES ARE THE RESULT OF
YEARS OF SCIENTIFIC STUDY COMBINED
WITH THE EXPERIENCE OF YEARS.

There are 6 Fs and 3 As in the sentence, and it is likely that in your class there will be different answers to this straightforward question (although most should get the correct number of As). There is no catch, but what happens is the human brain processes information in different ways and in some brains (but not all) we tend to see the F in OF as a V instead of an F.

Using this example with pupils explores the idea that something as basic as our reading isn't completely reliable, and teaches us that we all see things differently.

SEEING THINGS DIFFERENTLY (1)

SEEING THINGS DIFFERENTLY (2)

Although we often think that what we see is reality, the truth is that we can be consciously processing as little as 1/200,000 of what is actually happening, and as a result the brain tends to take shortcuts to overcome the filtering of information. An example of this worth sharing with pupils is the illusion below:

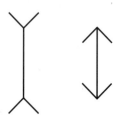

Everyone knows that the two vertical lines are exactly the same length (check them if you are not convinced) and that this is an optical illusion – more importantly, why does this happen? In this example, the brain, because it is working via shortcuts, is adding something to our perception: the apparently larger line represents an inner corner, while the shorter-appearing line represents an outer corner, and as most of us live in traditional rectilinear environments, it must as a consequence be further away. What is really interesting is that people brought up in non-rectilinear environments do not see the illusion. Even something as basic as what we see cannot always be trusted, and what we see others might not see. As they say, beauty is in the eye of the beholder, and when designing for others it is essential that we don't rely upon our own feelings too much!

Given that we have shown pupils that we all see things differently, it is important that they ultimately recognize this is why design is a contentious activity, and that design ideas are neither right nor wrong, but different to each viewer. Encourage pupils to be both critical of ideas while recognizing that ideas are merely opinions.

I was told about the design of a new building for severely disabled children with a whole range of physical and visual sensory problems. Many could barely walk or see.

When the new school was being built, the headteacher insisted the designers create the most uneven floor, with significant undulations and changes in texture every 10cm. I was quite surprised by this, and felt that the opposite might be more appropriate – perhaps thick carpet to protect the children when they fell. This was until it was explained to me: if they took my approach it would be better for the children in the short term, but in the long term it would only serve to institutionalize them. The most demanding learning environment would better prepare them for when they left the school.

I don't think the moral of this story needs too much explaining, and it is something we all need to consider. Are we doing what is best for our learners now, or are we preparing them effectively for the future?

LEARNING ENVIRONMENT

CHALLENGING DISADVANTAGE

If you sat down today and decided, from scratch, the most effective way of educating children and what they should learn, it is likely that it would look nothing like our current education system. What we have has slowly evolved over a relatively short period of a couple of hundred of years, and much of the practices that we have today have remained out of habit, and not because of good practice or educational rationale.

An example of this is the organization of the academic year. There is no real reason why we arbitrarily disadvantage those children born in the summer months (up until the end of August) by starting the academic year in September. Yet this one act disadvantages a significant number of children, as many are cognitively and emotionally up to a year behind their oldest peers.

Equally, the curriculum remains a combination of discrete, often out of date, subjects working in artificial and isolated ways that make little sense to many pupils. A feature in this book has been to suggest how D&T works best by developing learning dispositions that draw upon other learning rather than working in isolation.

Many of the ideas in this book are, in effect, trying to overcome some of the disadvantages that are naturally built into the education system and as successes of that system ourselves (teachers) we may not always recognize these disadvantages. It is therefore important that in deciding up the future we seek to remove the inbuilt disadvantages that exist, and to do this teachers need to consider what values, selective perceptions and preferences they may have which may inadvertently reinforce disadvantage.

So what do you take for granted – what has always been that way, yet might disadvantage some children – and most importantly what can you do about it?

There is nothing like a good quote to help capture the
moment, so here are a few favourites:

> 'Any intelligent fool can make things bigger and more
> complex.'
>
> Anon.

> 'Today's students are no longer the people our
> educational system was designed to teach.'
>
> Marc Prensky

> 'Your most unhappy customers are your greatest source
> of learning.'
>
> Bill Gates

> 'We need to rethink our ideas about what it means to
> be educated.'
>
> Ken Robinson

> 'I have never let my schooling interfere with my
> education.'
>
> Mark Twain

> 'The great aim of education is not knowledge but
> action.'
>
> Herbert Spencer

> 'Creativity is the opposite of copying.'
>
> Anon

> 'The illiterate of the twenty-first century will not be
> those who cannot read and write, but those who cannot
> learn, unlearn, and relearn.
>
> Alvin Toffler

> 'School is a place where children learn to be stupid!'
>
> John Holt

> 'It takes a village to raise a child.'
>
> Anon

> 'A teacher affects eternity; he can never tell where his
> influence stops.'
>
> Henry Brooks Adams